Methods in Molecular Biology

Series Editor
John M. Walker
School of Life and Medical Sciences
University of Hertfordshire
Hatfield, Hertfordshire, UK

For further volumes:
http://www.springer.com/series/7651

For over 35 years, biological scientists have come to rely on the research protocols and methodologies in the critically acclaimed *Methods in Molecular Biology* series. The series was the first to introduce the step-by-step protocols approach that has become the standard in all biomedical protocol publishing. Each protocol is provided in readily-reproducible step-by-step fashion, opening with an introductory overview, a list of the materials and reagents needed to complete the experiment, and followed by a detailed procedure that is supported with a helpful notes section offering tips and tricks of the trade as well as troubleshooting advice. These hallmark features were introduced by series editor Dr. John Walker and constitute the key ingredient in each and every volume of the *Methods in Molecular Biology* series. Tested and trusted, comprehensive and reliable, all protocols from the series are indexed in PubMed.

Genotyping

Methods and Protocols

Edited by

Anthony Torres

Cannabis Research Institute, Discovery Partners Institute, University of Illinois System, Chicago, IL, USA

Humana Press

Editor
Anthony Torres
Cannabis Research Institute, Discovery
Partners Institute
University of Illinois System
Chicago, IL, USA

ISSN 1064-3745 ISSN 1940-6029 (electronic)
Methods in Molecular Biology
ISBN 978-1-0716-4641-0 ISBN 978-1-0716-4642-7 (eBook)
https://doi.org/10.1007/978-1-0716-4642-7

This Humana imprint is published by the registered company Springer Science+Business Media, LLC, part of Springer Nature.
The registered company address is: 1 New York Plaza, New York, NY 10004, U.S.A.

Preface

Single nucleotide polymorphisms (SNPs) are the most common type of genetic variation. These single base-pair changes can have profound effects on an individual's susceptibility to disease, response to drugs, and other phenotypic traits. As such, SNP genotyping has become an essential tool for a wide range of applications, including plant marker-assisted breeding, medical diagnostics, pharmacogenomics, DNA fingerprinting, forensics, and population genetics.

This book offers a thorough overview of the various methods and protocols used for SNP genotyping. Our goal is to cover a wide range of techniques to accommodate different needs and levels of technological access. We explore both well-established methods such as Sanger sequencing and qPCR and cutting-edge approaches like nanopore sequencing and PACE. Each chapter delves into the principles behind each method, provides step-by-step protocols for implementation, and discusses their practical applications.

The chapters are organized thematically, beginning with foundational techniques and progressing toward more specialized applications. This edition of MiMB provides an overview of several important concepts and techniques related to SNP analysis. It begins with basic concepts, discussing SNP nomenclature, databases, and significance of SNPs across various fields. It then outlines traditional methods, detailing protocols for PCR-based techniques such as qPCR, HRM, and CD-PCR, along with classic approaches like Sanger sequencing and STR analysis. The text further explores isothermal amplification methodologies, highlighting innovative techniques like LAMP and RPA, which offer rapid and efficient alternatives to traditional PCR. Additionally, it covers advanced technologies, including high-throughput methods like nanopore sequencing, PACE, and FACS-based SNP genotyping. Finally, the document provides guidance on data analysis, emphasizing the importance of interpreting genotyping data with a focus on quality control, statistical analysis, and visualization.

Whether you are a seasoned researcher or a novice scientist entering the field, this book will serve as a valuable resource. We have strived to make the content accessible and engaging, with clear explanations, illustrative figures, and practical examples. Our goal is to empower readers with the knowledge and tools necessary to confidently perform genotyping and contribute to the advancement of this exciting field. We hope that this book will serve as a comprehensive guide, fostering a deeper understanding of genotyping and its myriad applications in the years to come.

Chicago, IL, USA *Anthony Torres*

Contents

Contributors

K. Aditya • *Division of Genomic Resources, ICAR—National Bureau of Agricultural Insect Resources, Bengaluru, India*

Jason Argyris • *Centre for Research in Agricultural Genomics (CRAG), CSIC-IRTA-UAB-UB, UAB, Bellaterra, Barcelona, Spain; IRTA (Institut de Recerca I Tecnologia Agroalimentaries), Barcelona, Spain*

Mark Arick II • *Institute for Genomics, Biocomputing and Biotechnology, Mississippi State University, Mississippi State, MS, USA*

P. Buella • *PathoGenOmics Lab, Department of Plant Pathology, University of Agricultural Sciences, Bengaluru, India*

Michael G. Campana • *Center for Conservation Genomics, Smithsonian National Zoo and Conservation Biology Institute, Washington, DC, USA*

Tašja Cvelbar • *Department of Biology, Biotechnical Faculty, University of Ljubljana, Ljubljana, Slovenia; Krka, d. d., Novo Mesto, Novo Mesto, Slovenia*

David Dobnik • *Department of Biotechnology and Systems Biology, National Institute of Biology, Ljubljana, Slovenia*

Reginald Gaudino • *Front Range Biosciences, Lafayette, CO, USA; Cannabis Research Institute, Discovery Partners Institute, University of Illinois System, Chicago, IL, USA*

Robert Givens • *Front Range Biosciences, Lafayette, CO, USA*

J. Harish • *PathoGenOmics Lab, Department of Plant Pathology, University of Agricultural Sciences, Bengaluru, India*

Chuan-Yu Hsu • *Institute for Genomics, Biocomputing and Biotechnology, Mississippi State University, Mississippi State, MS, USA*

Aisulu Ismailova • *Department of Information Systems, S. Seifullin Kazakh Agrotechnical Research University, Astana, Kazakhstan*

Amadej Jelenčič • *Department of Biotechnology and Systems Biology, National Institute of Biology, Ljubljana, Slovenia; Josef Stefan International Postgraduate School, Ljubljana, Slovenia*

Ruslan Kalendar • *National Laboratory Astana, Nazarbayev University, Astana, Kazakhstan*

R. Karan • *PathoGenOmics Lab, Department of Plant Pathology, University of Agricultural Sciences, Bengaluru, India*

Alexandra Bogožalec Košir • *Department of Biotechnology and Systems Biology, National Institute of Biology, Ljubljana, Slovenia*

Zenaida V. Magbanua • *Institute for Genomics, Biocomputing and Biotechnology, Mississippi State University, Mississippi State, MS, USA*

Jesús E. Maldonado • *Center for Conservation Genomics, Smithsonian National Zoo and Conservation Biology Institute, Washington, DC, USA*

C. Manjunatha • *Division of Genomic Resources, ICAR-National Bureau of Agricultural Insect Resources, Bengaluru, India*

Nataša Mehle • *Department of Biotechnology and Systems Biology, National Institute of Biology, Ljubljana, Slovenia; School for Viticulture and Enology, University of Nova Gorica, Vipava, Slovenia*

MOJCA MILAVEC • *Department of Biotechnology and Systems Biology, National Institute of Biology, Ljubljana, Slovenia*
YOICHI NAKANISHI • *Laboratory of Biochemistry, Graduate School of Bioagricultural Sciences, Nagoya University, Nagoya, Japan*
LILLIAN D. PARKER • *Laboratories of Molecular Anthropology and Microbiome Research, Department of Anthropology, University of Oklahoma, Norman, OK, USA; Center for Conservation Genomics, Smithsonian National Zoo and Conservation Biology Institute, Washington, DC, USA*
SWATHI S. PATIL • *PathoGenOmics Lab, Department of Plant Pathology, University of Agricultural Sciences, Bengaluru, India*
OLGA PECHANOVA • *Institute for Genomics, Biocomputing and Biotechnology, Mississippi State University, Mississippi State, MS, USA*
DANIEL G. PETERSON • *Institute for Genomics, Biocomputing and Biotechnology, Mississippi State University, Mississippi State, MS, USA*
D. PRAMESH • *Agricultural Research Station, Gangavathi, University of Agricultural Sciences, Raichur, India*
M. K. PRASANNAKUMAR • *PathoGenOmics Lab, Department of Plant Pathology, University of Agricultural Sciences, Bengaluru, India*
H. V. RAMEGOWDA • *Agricultural Research Station, Gangavathi, University of Agricultural Sciences, Raichur, India*
VLADISLAV SHEVTSOV • *Department of Information Systems, S. Seifullin Kazakh Agrotechnical Research University, Astana, Kazakhstan*
SHUNTA TAGA • *Laboratory of Biochemistry, Graduate School of Bioagricultural Sciences, Nagoya University, Nagoya, Japan*
ANTHONY TORRES • *Cannabis Research Institute, Discovery Partners Institute, University of Illinois System, Chicago, IL, USA*
NAOKI UNO • *Department of Medical Virology, Graduate School of Biomedical Sciences, Nagasaki University, Nagasaki, Japan*
GOPAL VENKATESHBABU • *PathoGenOmics Lab, Department of Plant Pathology, University of Agricultural Sciences, Bengaluru, India*

Chapter 1

An Assay for Variable Fragment Length Allele-Specific Genotyping

Ruslan Kalendar, Vladislav Shevtsov, and Aisulu Ismailova

Abstract

Polymerase chain reaction (PCR) is a quick and easy technique used to detect nucleotide polymorphisms and sequence variations in various fields such as medicine, agriculture, and basic research. There exists a set of PCR variants, collectively called allele-specific PCR, which use competitive reactions in the presence of allele-specific primers to amplify only specific alleles. In this chapter, we present protocols for a bioinformatics tool that can be used to design PCR-based genotyping assays for variable fragment length allele-specific genotyping. This assay is designed in both directions and can target single-nucleotide polymorphisms and insertion/deletions. This chapter provides a step-by-step guide to design a multiplexed primer assay for variable fragment length allele-specific genotyping.

Key words Single-nucleotide polymorphism (SNP), Insertion-deletion polymorphism (InDel), Genotyping system, Genotyping assay design software, Multiplex, SNaPshot, Multiplex Ligation-dependent Probe Amplification (MLPA)

Abbreviations

ASP	Allele-specific primer
AS-PCR	Allele-specific PCR
InDel	Insertion-deletion polymorphism
KASP	Kompetitive allele-specific PCR
SNP	Single-nucleotide polymorphism

1 Introduction

High-throughput technologies for nucleotide sequence analysis and detection of sequence variation are increasingly used for plant and animal genotyping, forensics, genetic medicine, and other areas of genetic testing [1]. The most common sequence variations in any genome are single-nucleotide polymorphisms (SNPs). The

Anthony Torres (ed.), *Genotyping: Methods and Protocols*, Methods in Molecular Biology, vol. 2943,
https://doi.org/10.1007/978-1-0716-4642-7_1,

primary consideration in selecting a marker type for genotyping is the polymorphisms' information content and the test's power. In this regard, SNPs are potent markers [2, 3]. Today, there is an enormous number and variety of established SNP analysis and genotyping methods based on very different platforms and approaches [4–8].

PCR can be adapted for the rapid detection of single base changes in genomic DNA using a family of closely related methods, such as allele-specific PCR (AS-PCR), PCR amplification of specific alleles (PASA), allele-specific amplification (ASA), and amplification refractory mutation system (ARMS). All these methods use specifically designed PCR primer sets containing allele-specific primers (ASP), in which allele specificity is determined by the base at or near the 3′ end. In principle, AS-PCR assays can be designed to analyze almost any allelic variation [4]. Other improvements in AS-PCR include fluorescence to detect the amplification of a specific allele. Singleplex PCRs for SNP analysis, such as TaqMan (Life Technologies, USA) and SimpleProbe (Roche Applied Science, USA), have been described as valuable additions to marker-assisted selection in genetic medicine and other areas of genetic testing.

Kompetitive Allele-Specific PCR (KASP) (LGC Biosearch Technologies, United Kingdom), PCR Allele Competitive Extension (PACE) (3CR Bioscience, United Kingdom), and rhAmp (Integrated DNA Technologies, USA) genotyping systems are developments of the original AS-PCR approach. These techniques use fluorescence resonance energy transfer (FRET) for signal generation, and both allow accurate bi-allelic discrimination of known SNPs and insertion-deletion polymorphisms (InDels) [9]. FRET is a quantum phenomenon that occurs between two fluorophore molecules [10]. Excitation is transferred from a donor fluorophore to an acceptor fluorophore by dipole–dipole interaction without the emission of a photon. This quenches the fluorescence of the donor molecule and excites the acceptor molecule.

In KASP, PCR amplification uses a pair of allele-specific (forward) primers and a single common (reverse) primer. In addition to the common primers, a FRET cassette is added to the reaction mixture. This is a duplex of two synthetic complementary oligonucleotides, one labeled with a fluorescent dye and the other carrying a fluorescence quencher. In addition, each allele-specific primer (ASP) has a unique 5′ terminal extension (tail) complementary to the FRET cassette sequence. During amplification of the allele involving the tailed ASP, there is an increase in the amount of DNA to which the dye-labeled moiety anneals. This disrupts the integrity of the FRET cassette, and the fluorescent dye is spatially separated from the quencher, allowing it to emit fluorescence. Unlike other PCR-based genotyping assays, KASP does not require labeling of the target-specific primers/probes, providing additional flexibility in assay design. KASP technology is particularly suitable for high-volume screening projects like plant breeding. A key feature of KASP technology is a universal FRET cassette reporter system, eliminating the need for double-labeled probes. Commercial companies produce PCR additives, called master mixes, which contain one or more "universal" FRET cassette(s).

Allele-specific quantitative PCR (ASQ), like KASP and other techniques, is based on the simultaneous presence in the PCR of the following two components: an allele-specific mixture (allele-specific and common primers) and a template-independent detector mixture containing two or more (up to four) universal probes and a single universal quencher oligonucleotide (Fig. 1) [11, 12]. It can be used for 3- or 4-allelic variants or different SNPs in a multiplexed format in several SNP and InDel genotyping applications. In these approaches, allele discrimination is performed using different fluorescent dyes, and the method is suitable for widespread use in real-time PCR [13–15].

Specialized PCR techniques for SNP and InDel genotyping have also been proposed using discrimination of each allele by PCR product size [16, 17]. These PCR techniques for variable fragment length allele-specific genotyping are, for example, the SNaPshot, amplification refractory mutation system PCR (ARMS-PCR), PCR restriction fragment length polymorphism (PCR-RFLP), and cleaved amplified polymorphic sequence (CAPS) [18]. However, there is also a way to discriminate alleles using amplicon size variation. Any existing approach can be converted to differentiate allelic variants by amplicon length. For instance, for the KASP technology, instead of using a FRET cassette, it is sufficient to use a non-specific oligo 5′ tail of different lengths to allow simultaneous identification of each allele by gel or capillary electrophoresis.

The SNaPshot Multiplex System (Thermo Fisher Scientific, USA) is a primer extension-based method developed for the analysis of single-nucleotide polymorphisms (SNPs) (Fig. 2). The SNaPshot technique assays up to ten SNP markers simultaneously using PCR amplification followed by dideoxy single-base extension of an unlabeled primer [19–23]. The primer is designed to anneal to the sequence adjacent to the SNP site. Once the primer has annealed, a single-base extension occurs by adding the complementary dye-labeled ddNTP (dye terminator) to the annealed primer. Each of the four ddNTPs is fluorescently labeled with a different colored dye. The result is marker fragments for the other SNP alleles, all the same length but differentiated by color. After electrophoresis and fluorescence detection, the alleles of a single marker appear as different colored peaks of approximately the same size in the electropherogram plot. The size of the different allele peaks will vary slightly due to differences in the molecular weight of the dyes. The GeneMapper software is then used for sizing and genotyping.

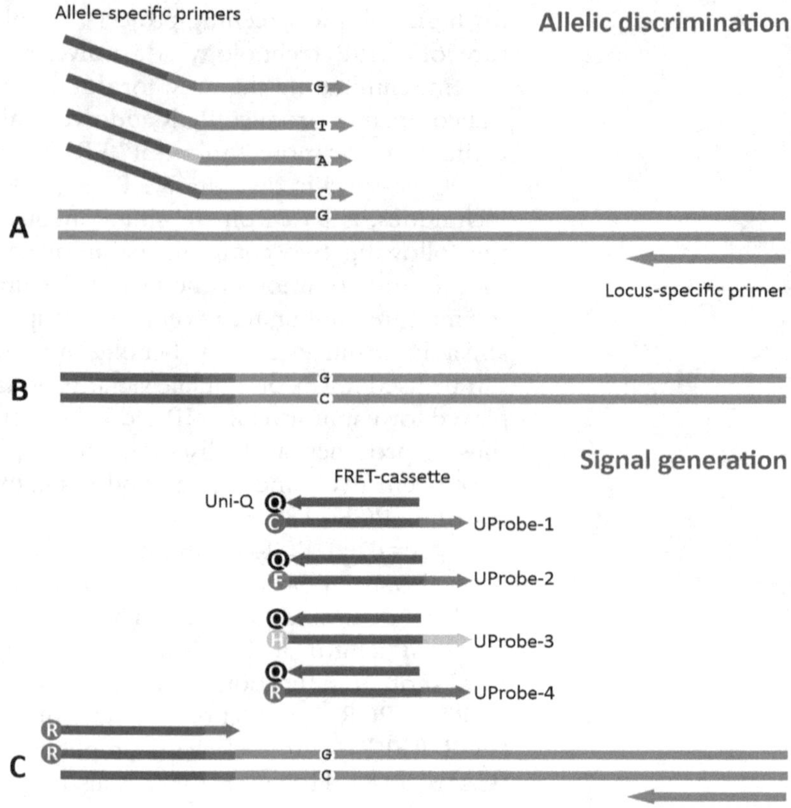

Fig. 1 Four-plex fluorescent ASQ AS-PCR assay genotyping system. For ASQ, the FRET cassette consists of two or more allele-specific primers (ASP) and a fluorescently labeled universal probe (UP) with a single universal quencher probe (Uni-Q). Differences in the tail sequence ASPs and UPs are determined by a unique 6-nt barcode sequence that is not part of the universal tail of the Uni-Q sequence. The design of an allele-specific primer with the SNP site at the penultimate or antepenultimate 3′ base of each allele-specific primer is characteristic of the ASQ method. (**a**) Allele-specific primers query the SNP locus. Denaturation of DNA template and annealing ASP to the target, PCR round 1. (**b**) Formation of a PCR product containing a specific tail sequence complementary to allele-specific primers. This PCR product will be used in subsequent PCR cycles as a template for amplification using a specific fluorescently labeled UP (**c**). During the first two amplification cycles, a tail sequence is incorporated into the amplicon and subsequently recognized by a universal, probe-based reporter system

Multiplex ligation-dependent probe amplification (MLPA) is a simple, high-throughput, and easy-to-use method for detecting DNA targets of up to 40 sequences in a single reaction (Fig. 3). This technology (MRC Holland, Netherlands) was initially developed for the study of disease-associated gene copy number variations (CNVs) [24–27]. MLPA is also used to detect DNA methylation changes (MS-MLPA) and is sensitive enough to

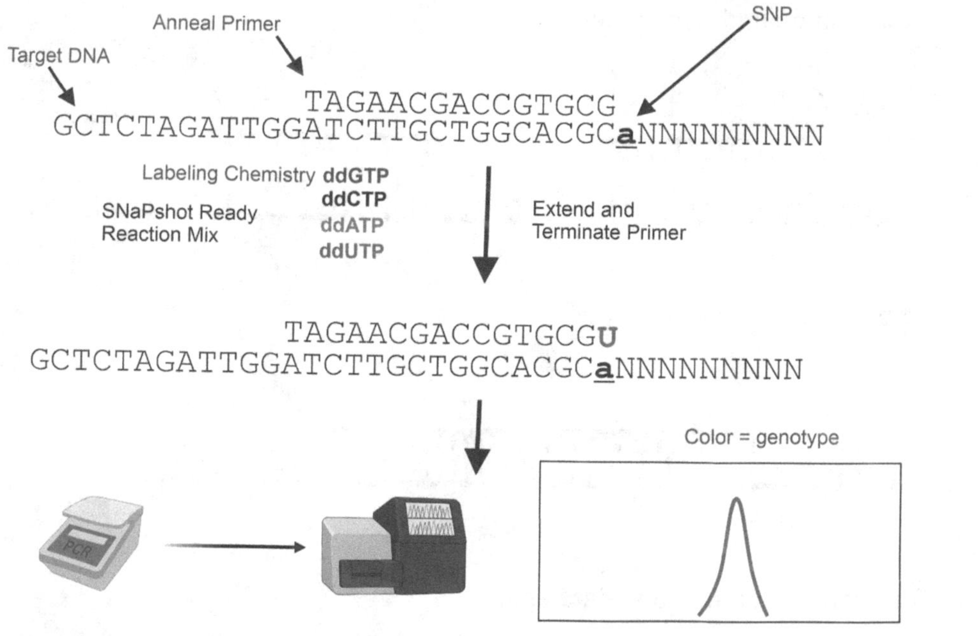

2 The SNaPshot labeling chemistry is based on single-base extension and termination. The chemistry is on a dideoxy single-base extension of an unlabeled primer. Each primer binds to a complementary in the presence of fluorescently labeled ddNTPs and DNA polymerase. The polymerase extends the ne nucleotide and adds a single ddNTP to its 3′ end—the fragment separation by capillary The fluorescent color readout indicates which base has been added

distinguish changes in disease-causing genes from highly similar pseudogenes. However, the MLPA technique is ideal for distinguishing SNP variants or InDel polymorphisms, if each individual 'lele is encoded in a unique probe [28]. This relatively simple nique is based on the semi-quantitative polymerase chain reac- inciple. It can detect allele-specific changes, quantify geno- methylation, or perform mRNA profiling. MLPA probes wo oligonucleotides, each containing a PCR primer a sequence complementary to the target, called the uence. The two probes hybridize immediately ther. When the probes hybridize correctly to they are ligated by a thermostable ligase s around the ligation site are allowed, on highly specific. The PCR primers ated probes. One of the primers is visualize the amplification product on assays can be run in a single d not be removed. Sequencing ate the resulting PCR products. designed to be easily identified when oduct is run through a gel. The filler ifference in size.

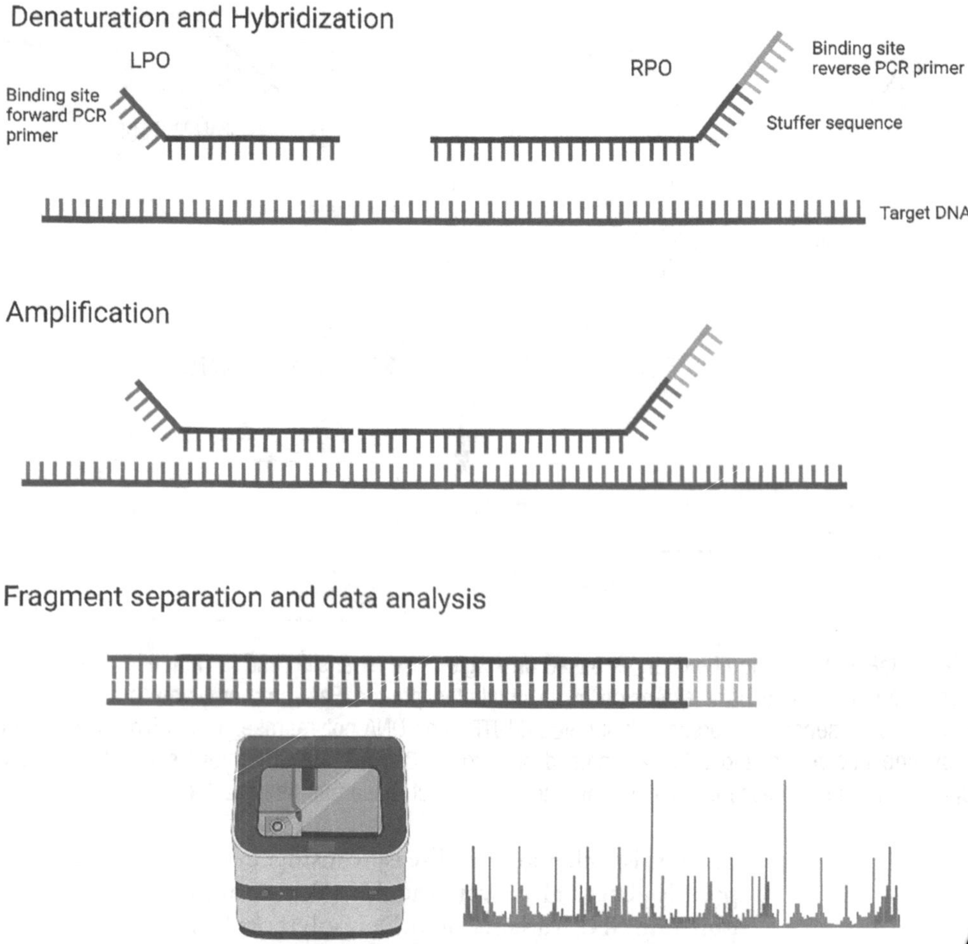

Fig. 3 Multiplex ligation-dependent probe amplification (MLPA) uses pairs of probes to detect specific s of the genome. MLPA probemixes have probes that target a specific genomic sequence. An ML consists of a left and a right probe oligonucleotide (LPO and RPO). The LPO and RPO contain PCR DNA hybridization sequences. An additional stuffer sequence in the RPO gives each probe a uni LPO and RPO parts of each probe hybridize to immediately adjacent target DNA sequences. The l probe oligos that have hybridized to immediately adjacent target sequences. The number of products measures the number of target sequences present in the sample. The ligated probes a multiplex PCR using a single universal primer pair. Only ligated probes are amplifie eliminating the need to remove unbound and unligated probes—the fragment separa electrophoresis. PCR products are loaded onto a capillary electrophoresis device and se Each fragment corresponds to a specific MLPA probe

Competitive allele-specific TaqMan-MGB (Thermo Fisher Scientific, USA) is a modificat based allele-specific PCR, which works by allele-specific primer (ASP) with a wild-type (ASB) in the same PCR reaction. Usin completely suppresses the amplification o

as not to interfere with the amplification of the mutant allele and improve the specificity and sensitivity of the assay [29].

The Invader assay uses a structure-specific flap endonuclease (FEN) to cleave a three-dimensional complex formed by hybridizing allele-specific overlapping oligonucleotides to target DNA containing an SNP site [30]. Annealing of the oligonucleotide complementary to the SNP allele to the target molecule triggers cleavage of the oligonucleotide by Cleavase, a thermostable FEN enzyme. Cleavage can be detected using several different approaches. The cleavage product commonly triggers a secondary cleavage reaction on a FRET cassette to release a fluorescent signal. Alternatively, cleavage can be detected directly using fluorescence polarization (FP) probes or mass spectrometry. The invasive cleavage reaction is particular, has a low failure rate, and can detect femtomolar amounts of target DNA. While the assay has traditionally been used to interrogate one SNP in one sample per reaction, novel chip- or bead-based approaches are being tested to make this efficient and accurate assay adaptable to multiplexing and high-throughput SNP genotyping.

The tools described here are integrated into and operate within the FastPCR environment [31–33]. The FastPCR program, including the many AS-PCR design tools, is presented here. This allele-specific genotyping tool simplifies and automates the design of genotyping assays, resulting in a higher probability of success.

2 Software

The FastPCR software (https://primerdigital.com/fastpcr.html) can be used with any version of Microsoft Windows. The online version of FastPCR software (https://primerdigital.com/tools/kasp.html) is implemented in JavaScript to ensure universal compatibility across all operating systems.

3 Methods

3.1 Basic Principle of the Algorithm for the Design of Allele-Specific Primers

The ASQ genotyping approach enables the detection of up to four spectrally separated dyes (Fig. 1). The FastPCR software automatically selects the primer length and the position of its 3′ terminus for each ASP. This type of optimization is also necessary when working with InDels. For example, in the case of a null site polymorphism (i.e., one or more nucleotide(s) are missing in one allele, or there are insertions in other alleles), the ASP sequence targeting that allele is positioned so that the 3′-terminal sequences of the primer are sufficient for annealing. In this way, the ASP sequences are correctly adapted for each allelic variant. In the case of short InDels (<4 nucleotides), the better solution is to design the ASP 3′ termini sequence individually for each variant, which can be achieved with FastPCR. Thus, the algorithm allows the generation of ASPs with 3′-terminal sequences unique for each allelic variant. For InDels, the potential for 3′ end generation is more significant for each ASP design option, offering the possibility of more effective discrimination of the various alleles. The software also determines the minimum length of the 3′ end of each ASP for effectively detecting all polymorphisms. The user can define the position of the polymorphic site at any position, such as a first, second, or third base in the 3′ end of each ASP, to increase reaction specificity and allele discrimination. The 3′ end sequence of each ASP should be designed so that each ASP is unique and specific only to its target. The software uses the sequence from the insertion to generate the 3′ end of the ASP sequence for a null site polymorphism. In the case of a single or double base InDel, the following sequence is used after the target site to generate the 3′ end of each ASP. This should be sufficient for each InDel polymorphism to generate unique sequences for all 3′ ends of ASPs. Primer sequences are optimized for similar thermodynamic parameters (Tm, Gibbs free energy (dG)). To match the thermodynamic parameters of each ASP, the program extends the length of the low Tm primer sequence to approximate the Tm of a primer with more G/C groups at the 3′ end.

3.2 Interface

The software contains menus, toolbars, a ribbon, and three text editors. The ribbon is designed to help the user quickly find the commands needed to complete a task. Commands are organized into logical groups, which are grouped into tabs. Each tab corresponds to a type of activity, such as “PCR primer design,” “in silico PCR,” or “primer testing,” and so on.

There are three independent text editors under different tabs: “General Sequence(s),” “Additional Sequence(s) or Prepared Primers (Probes) List,” and “Result Report.”

The first two text editors are necessary for loading sequences for analysis; the General Sequence(s) text editor is designed for working with project sequences. The “Additional sequence(s) or pre-designed primers (probes) list” text editor can be used for unique and additional sequences, such as pre-designed primers, multiple query sequences, or the numbers to be entered.

3.3 Sequence Entry

Getting started with a basic project in FastPCR software is as easy as opening a new or existing text file, using copy-paste, or just typing. The sequence data file is prepared using a text editor and saved in ASCII text/plain format. The software accepts a single sequence or multiple separate DNA sequences in FASTA or tab-delimited text file format. The FASTA format is preferred; it is simply the raw sequence preceded by a definition line. The definition line begins with a “>” character, optionally followed by a sequence name of any length and number of words, with no spaces in between. Many sequences can be listed in the same file. The format requires that a new sequence always starts with a new ">" symbol. Template length is not limited. The FastPCR clipboard allows users to copy and paste text or tables from Microsoft Office documents, an Excel spreadsheet, or other programs into another Office document. All target sequences must be prepared in the same format. The user can type or import from file(s) into the "General Sequence(s)" or "Additional Sequence(s) or Prepared Primers (Probes) List" editors.

Users can design compatible primer sets for a given primer (probe) or a list of given primers (probes). This feature can be helpful when planning a multiplex assay for different targets. FastPCR allows users to avoid non-specific amplification by selecting the best-paired primer for a given oligonucleotide to achieve the highest probability of success.

Degenerate DNA sequences are accepted as IUPAC code, an extended vocabulary of 11 letters that allows the description of ambiguous DNA code. Each letter represents a combination of one or more nucleotides: M = (A/C), R = (A/G), W = (A/T), S = (G/C), Y = (C/T), K = (G/T), V = (A/G/C), H = (A/C/T), D = (A/G/T), B = (C/G/T), N = (A/G/C/T).

3.4 Specifying a Polymorphic Site to the Software

Examples of the input of sequences encoding polymorphic sites can be found in the file menu of the FastPCR software. Input sequences can be in FASTA, tabular formats, or multiple sequence alignments, as shown in the examples. Square brackets indicate the boundaries of a polymorphic site of interest. Brackets indicate the alternative alleles of an SNP, e.g., [C/G] (alternative coding—[S]). The software imposes no size limit on the length difference for InDels alleles, provided all alleles can be aligned and the alignment has a sufficient overlap to target the primers. Two, three, or four allelic variants can be included in the analysis to calculate allele-specific primers (ASPs) and a common primer. Only the differences between the variants need to be shown in parentheses. An allele-specific primer is designed for each input allelic variant. A common primer targeting a conserved region in all sequences is also generated. Individual allelic variants in a polymorphic site are thus listed as [variant1/variant2] for binary polymorphism or [variant1/variant2/variant3/variant4] for quaternary polymorphism. Alternatively, IUPAC notation can be used for a degenerate base (e.g., FASTA format). If a polymorphic site contains InDels, the same format with square brackets ([variant1/variant2/variant3/variant4]; two or more different variants per site) describes individual variants. Only the differences between the variants need to be specified within the brackets.

3.5 Selecting a Specific Task Command for a Particular Application

The software allows the design of primers with the SNP placed at any position concerning the 3′ end. To do this, the user must specify the desired SNP position using the position number command. By default, the program automatically sets the SNP position to the penultimate base (-**aspcr**). In addition, for InDel sites, the number of discreditable nucleotides at the 3′ end of each ASP can be automatically determined depending on the polymorphism variant.

3.6 Adding Non-template DNA Sequences to Primer Ends

The tail sequence can optionally be specified up to the 5′ end with the command '-**5e[NN]**', where "NN" is a sequence of one or more bases, e.g., '-**F5e**[CGACG] -**R5e**[TTTTTTTTTTTTTTT]', adds sequence 'CGACG' to forward primers and sequence 'TTTTTTTTTTTTTTT' to reverse primers at 5′ ends.

For tasks requiring multiple allelic variants with individual added tail sequences that a separator can separate, the slash character "/" is used to denote each tail variant. For example, for bi-allelic scoring of SNPs and indels at specific loci related to the development of competitive AS-PCR genotyping assays, two (up to four) variants of the added tail sequences are as follows:

-**F5e**[gaaggtgaccaagttcatgct/gaaggtcggagtcaacggatt].

3.7 Allele-Specific PCR Assay Design

In this example, an allele-specific PCR is designed to genotype SNP alleles (Fig. 4). The SNP of interest must be enclosed in [square brackets] and can be formatted as [first allele/second allele/third allele/fourth allele] or [IUPAC code] or [target nucleotide]:

>1

tcatattccagtttgggcgagttttaagataggtccgg [C/G] acagtctttgcggcgccaacgcgtctttctccagcagacagtccccggactgc

>2

tcatattccagtttgggcgagttttaagataggtccgg [S] acagtctttgcggcgccaacgcgtctttctccagcagacagtccccggactgc

>3

tcatattccagtttgggcgagttttaagataggtccgg [C] acagtctttgcggcgccaacgcgtctttctccagcagacagtccccggactgc

AS-PCR assays can be designed to discriminate InDels polymorphisms:

>1

tcatattccagtttgggcgagttttaagataggtccgg [AG/-] acagtctttgcggcgccaacgcgtctttctccagcagacagtccccggactgc

>2

tcatattccagtttgggcgagttttaagataggtccgg [AG/TT] acagtctttgcggcgccaacgcgtctttctccagcagacagtccccggactgc

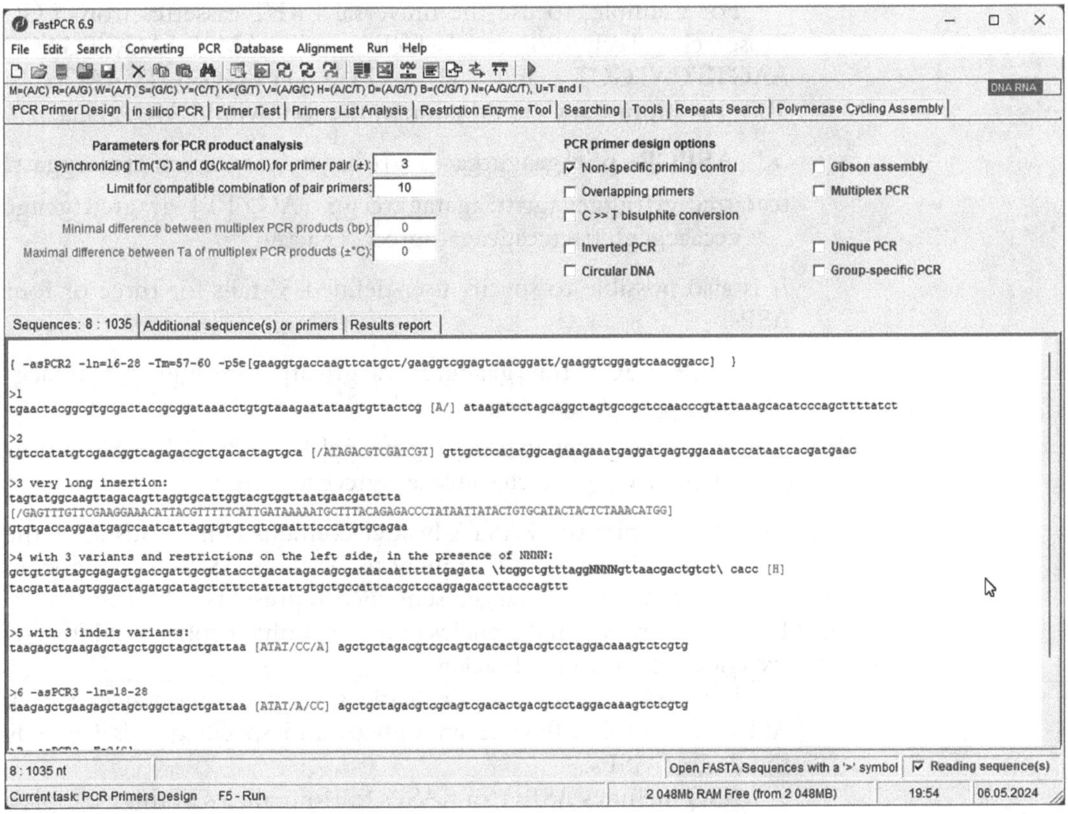

Fig. 4 FastPCR program interface with examples for developing allele-specific polymerase chain reaction genotyping assays. Several generic examples of SNP and InDel genotyping are shown. Also shown are the command (-asPCR2 -ln=16-28 -Tm=57-60 -p5e[gaaggtgtgaccaagttcatgct/gaaggtgtcggtcggagtcaacgggatt/gaaggtcggtcggtcaacggacc]) to indicate to the program what length primers should be (-ln=16-28), in what range the melting temperature of primers should be (-Tm=57-60) and three variants of tailed sequences for SNP variants

To instruct the software to design an AS-PCR assay, a user must place the "-**ASPCRn**" command in the FASTA header of a specific sequence or globally within curly brackets. The command can instruct the software to position the first, second, or third base at the 3′ terminus of the ASP above the SNP site. To do this, replace "n" in this command with a number (1…4). The default is -**ASPCR2** (which means that a variable base is at the second to last position at the 3′ end of the ASP).

If the user does not specify the number: -**ASPCR**, the program will automatically determine the optimal length for effective discrimination of single-nucleotide polymorphisms.

If it is necessary to add user-defined tails to the 5′-termini of ASPs, this can be done by defining a specific 5′-tail for each ASP. To do this, a tail sequence must be included in the FASTA header after the "-p5e" command, and individual tail sequences must be separated by a slash "/."

For example, to use the universal FRET cassettes from LGC Biosearch Technologies (which are FAM-GAAGGTGACC AAGTTCATGCT and HEX-GAAGGTCGGAGTCAACGGA TT), the input data should contain the following FASTA header:

>1 -**ASPCR** -**p5e**[gaaggtgaccaagttcatgct/gaaggtcggagtcaacggatt]

tcatattccagtttgggcgagttttaagataggtccgg [AG/TT] acagtctttgcggc gccaacgcgtctttctccagcagacagtccccggactgc

It is also possible to specify user-defined 5′-tails for three or four ASPs:

>1 -**ASPCR** -**p5e**[gaaggtgaccaagttcatgct/gaaggtcggagtcaacg-gatt/gaaggtcggagtcaacggacc]

tcatattccagtttgggcgagttttaagataggtccgg[ATAT/CC/A]acagtctttgc ggcgccaacgcgtctttctccagcagacagtccccggactgc

In this example, the FASTA header contains commands after the ">" sign and a list of tail sequences for three ASPs. Starting from the following line, a target sequence represents all three alleles. Differences between the alleles (e.g., the polymorphism of interest) are enclosed in square brackets.

It is not necessary to explicitly specify tail sequences for all ASPs. The tool is flexible enough to add specified tails for only the desired ASPs.

ASQ includes ASPs (four ASPs for four alleles in this example), a common universal primer (UP), and a detector system. The detector system comprises universal probes (ccagctgaacggtGCT-CAC/ccagctgaacggtTGCGAC/ccagctgaacggtGTGCGA/ccagct-gaacggtGCGTCA) and a universal quencher oligonucleotide (5′-accgttcagctgg). The FastPCR input is used to design the four-plex ASQ:

-**ASPCR2** -**5e**[ccagctgaacggtGCTCAC/ccagctgaacggtTGCGAC/ ccagctgaacggtGTGCGA/ccagctgaacggtGCGTCA]

>1

cttagatcgacaggtctaagagctgaagagctagctattaaagtcgagc[N]agctgctaga cgtcgcagtcgacacagctagcctaggacaaagtctcgtg

The target sequence contains a SNP polymorphism with four states: [N]. AS-PCR command: -ASPCR2 for SNP located at the penultimate base in the 3′ end of each allele-specific primer (ASP).

3.8 SNaPshot Assay Design

SNaPshot is an example of a primer extension-based method for genotyping known SNP positions using Applied Biosystems automated DNA analyzer. Its multiplexing capability allows up to ten SNPs to be analyzed in a single reaction using unlabeled, custom primers. To perform the SNaPshot task, use the command (-**snapshot**) in front of the list of target sequences in braces. For example, as in this case, after the list of target sequences, square brackets are used to specify the boundaries of a polymorphic site of interest (alternate encoding—[S]).

>1 **-snapshot** -pTm=55-58 -pln=16-35 -popyes (100-200).

agagcctacttgcataatccatttagtaagtcggataattaaggtgccggtatacgt[S] gtaacttaaacggaacgccaacgaaaatcaattagcttgactgtgggagtc

3.9 Multiplex Ligation-Dependent Probe Amplification (MLPA) Assay Design

MLPA or a digital MLA (dMLA) example for mutation detection (SNPs/InDels), up to 40 sequences in a single reaction. To perform the MLPA task, use the command (**-mlpa**) in front of the list of all target sequences in curly brackets. For example, as in this case:

>1 **-mlpa** -P5e[GGGTTCCCTAAGGGTTGGA/ TCTAGATTGGATCTTGCTGGCAC]

agagcctacttgcataatccatttagtaagtcggataattaaggtgccggtatacgt[S] gtaacttaaacggaacgccaacgaaaatcaattagcttgactgtgggagtc

3.10 Performing a Selected Task

Select PCR -> PCR Options from the menu bar for an AS-PCR analysis. The PCR Primer or Probe Design Options dialog box is displayed. The PCR Primer or Probe Design Options dialog allows the user to set basic parameters for the PCR reaction and the primers that will be generated. The PCR Primer or Probe Design Options dialog box provides several execution options for selecting the PCR type and key PCR parameters. By default, the program selects primer sets with compatible melting temperatures. Alternatively, the user can choose primer sets with similar dG.

The user selects the ribbon with a desired task (PCR Primer Design). The program performs only the chosen task. Depending on the user's chosen task, pressing **F5** will execute the current task. The result is displayed in the Results Report text editor when the task is complete (Fig. 5).

3.11 Program Output

FastPCR automatically generates results in tabular format in the third text editor, "Results Report."

Results with program-suggested primers and primer sets can be exported in standard exchange formats (e.g., MS Excel). Alternatively, the results can be easily saved as a .xls or .rtf text file compatible with Excel or OpenOffice. The separate output of the primer design is a list of primers, a set of primer pair sequences with their theoretical PCR products. The following parameters can be entered into spreadsheets: primer name, sequence, position on target, sequence direction, length, melting temperature, GC content, molecular weight, molar extinction coefficient, length complexity (LC), and primer quality (PQ). The tool checks all primers for self-dimer and cross-dimer interactions. The software selects primers that avoid repeats or other stable motifs, such as potential G-quadruplex sequences.

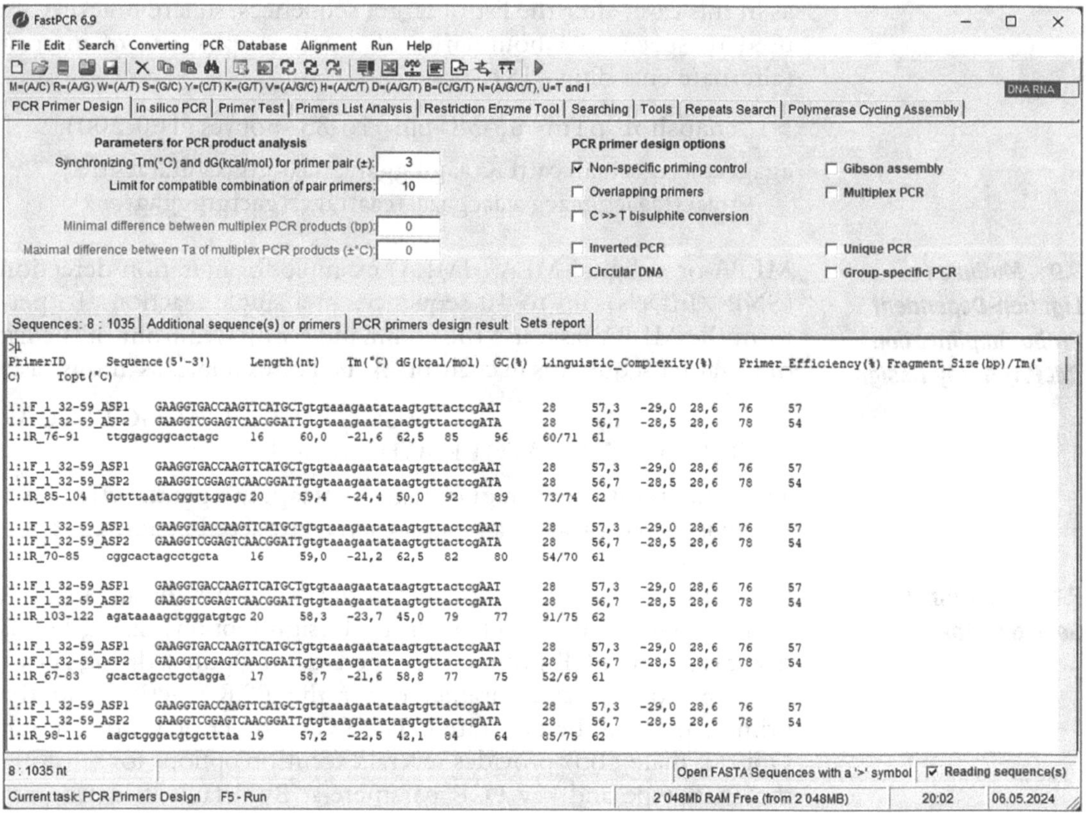

Fig. 5 The result obtained from the developed allele-specific polymerase chain reaction genotyping assays. Each set is separated by a space from the other variants and includes allele-specific primers and a common primer. The allele-specific primers contain unique nonspecific tails (highlighted with a capital letter) at their 5-termini and contain unique allele-specific sequences (highlighted with a capital letter) at their 3-termini

The recommended annealing temperature, product size (in bp), and Tm of a PCR product are provided for primer pairs. The program calculates several primer pairs or sets for the user to select the best. The user can use the most appropriate set of primers from the output, or if desired, alternative sets of primers can be used to target the same site. The user can request a desired product size or instruct the software to search for primer pairs for diagnostic PCR; in this case, an entire input sequence is used to search for the best primer pair.

4 Conclusion

This chapter describes a software tool in the FastPCR environment to develop PCR-based genotyping assays. The tool assists in the design of assays for detecting SNPs and InDels. FastPCR expands detection capabilities by allowing the detection of all four variants at a single SNP site and any form of InDel. Theoretically, it also allows the genotyping of several different SNP targets in a multiplexed reaction and the genotyping of mixed polymorphisms. The AS-PCR design tool can be used to plan high-throughput screening assays to rapidly identify mutant strains of microorganisms, specific genetic lines of higher organisms, or cell lines, such as those produced by genome editing using transcription activator-like effector nucleases (TALENs) or clustered regularly interspaced short palindromic repeats/CRISPR-associated protein (CRISPR/Cas). Another entity presented in this chapter is a multiplexed primer assay for variable fragment length allele-specific genotyping for quantitative measurement of the presence of allelic variants.

Acknowledgments

This research has been funded by the Committee of Science of the Ministry of Science and Higher Education of the Republic of Kazakhstan (Grant No. AP19678041 and BR24993023, BR24992881, BR27199879, BR24992841).

Conflict of Interest

The authors have no conflicts of interest to declare that are relevant to the content of this chapter.

References

1. Giner-Delgado C, Villatoro S, Lerga-Jaso J, Gaya-Vidal M, Oliva M, Castellano D, Pantano L, Bitarello BD, Izquierdo D, Noguera I, Olalde I, Delprat A, Blancher A, Lalueza-Fox C, Esko T, O'Reilly PF, Andres AM, Ferretti L, Puig M, Caceres M (2019) Evolutionary and functional impact of common polymorphic inversions in the human genome. Nat Commun 10(1):4222. https://doi.org/10.1038/s41467-019-12173-x
2. Arvas YE, Marakli S, Kaya Y, Kalendar R (2023) The power of retrotransposons in high-throughput genotyping and sequencing. Front Plant Sci 14:1174339. https://doi.org/10.3389/fpls.2023.1174339
3. Long YM, Chao WS, Ma GJ, Xu SS, Qi LL (2017) An innovative SNP genotyping method adapting to multiple platforms and throughputs. Theor Appl Genet 130(3):597–607. https://doi.org/10.1007/s00122-016-2838-4
4. Lawrie RD, Massey SE (2023) Agrigenomic diversity unleashed: current single nucleotide polymorphism genotyping methods for the agricultural sciences. Appl Biosci 2(4):565–585. https://doi.org/10.3390/applbiosci2040036
5. Kalendar R, Shustov AV, Schulman AH (2021) Palindromic sequence-targeted (PST) PCR, version 2: an advanced method for high-throughput targeted gene characterization and transposon display. Front Plant Sci 12:691940. https://doi.org/10.3389/fpls.2021.691940
6. He Q, Hu O, Chen M, Liang Z, Liang L, Chen Z (2022) A novel and cost-efficient allele-specific PCR method for multiple SNP genotyping in a single run. Anal Chim Acta 1229:340366. https://doi.org/10.1016/j.aca.2022.340366
7. Li B, Liu Y, Hao X, Dong J, Chen L, Li H, Wu W, Liu Y, Wang J, Wang Y, Li P (2021) Universal probe-based intermediate primer-

triggered qPCR (UPIP-qPCR) for SNP genotyping. BMC Genomics 22(1):850. https://doi.org/10.1186/s12864-021-08148-2

8. Touroutine D, Tanis JE (2020) A rapid, superselective method for detection of single nucleotide variants in Caenorhabditis elegans. Genetics 216(2):343–352. https://doi.org/10.1534/genetics.120.303553
9. Jatayev S, Kurishbayev A, Zotova L, Khasanova G, Serikbay D, Zhubatkanov A, Botayeva M, Zhumalin A, Turbekova A, Soole K, Langridge P, Shavrukov Y (2017) Advantages of Amplifluor-like SNP markers over KASP in plant genotyping. BMC Plant Biol 17(Suppl 2):254. https://doi.org/10.1186/s12870-017-1197-x
10. Förster T (2006) Zwischenmolekulare Energiewanderung und Fluoreszenz. Ann Phys 437(1–2):55–75. https://doi.org/10.1002/andp.19484370105
11. Kalendar R, Baidyussen A, Serikbay D, Zotova L, Khassanova G, Kuzbakova M, Jatayev S, Hu YG, Schramm C, Anderson PA, Jenkins CLD, Soole KL, Shavrukov Y (2021) Modified "allele-specific qPCR" method for SNP genotyping based on FRET. Front Plant Sci 12:747886. https://doi.org/10.3389/fpls.2021.747886
12. Kalendar R, Shustov AV, Akhmetollayev I, Kairov U (2022) Designing allele-specific competitive-extension PCR-based assays for high-throughput genotyping and gene characterization. Front Mol Biosci 9:773956. https://doi.org/10.3389/fmolb.2022.773956
13. Lee HB, Schwab TL, Koleilat A, Ata H, Daby CL, Cervera RL, McNulty MS, Bostwick HS, Clark KJ (2016) Allele-specific quantitative PCR for accurate, rapid, and cost-effective genotyping. Hum Gene Ther 27(6):425–435. https://doi.org/10.1089/hum.2016.011
14. Toth T, Csaba A, Bokor A, Acs N (2023) Variable fragment length allele-specific polymerase chain reaction (VFLASP), a method for simple and reliable genotyping. Mol Cell Probes 69:101910. https://doi.org/10.1016/j.mcp.2023.101910
15. Unsal SG, Yeni O, Buyuk U, Ozden Ciftci Y (2024) A novel method of multiplex SNP genotyping assay through variable fragment length allele-specific polymerase chain reaction: multiplex VFLASP-ARMS. Mol Cell Probes 75:101960. https://doi.org/10.1016/j.mcp.2024.101960
16. Tortajada-Genaro LA, Puchades R, Maquieira A (2017) Primer design for SNP genotyping based on allele-specific amplification—application to organ transplantation pharmacogenomics. J Pharm Biomed Anal 136:14–21. https://doi.org/10.1016/j.jpba.2016.12.030
17. Kalendar R, Shustov AV, Seppanen MM, Schulman AH, Stoddard FL (2019) Palindromic sequence-targeted (PST) PCR: a rapid and efficient method for high-throughput gene characterization and genome walking. Sci Rep 9(1):17707. https://doi.org/10.1038/s41598-019-54168-0
18. Shirasu N, Yasunaga S (2016) Duplex PCR-RFLP for the simultaneous genotyping of single nucleotide polymorphisms in ADH1B and ALDH2 genes. Anal Sci 32(12):1363–1366. https://doi.org/10.2116/analsci.32.1363
19. Zar MS, Shahid AA, Shahzad MS, Shin KJ, Lee HY, Lee SS, Israr M, Wiegand P, Kulstein G (2018) Forensic SNP genotyping with SNaPshot: development of a novel in-house SBE multiplex SNP assay. J Forensic Sci 63(6):1824–1829. https://doi.org/10.1111/1556-4029.13783
20. Daniel R, Santos C, Phillips C, Fondevila M, van Oorschot RA, Carracedo A, Lareu MV, McNevin D (2015) A SNaPshot of next generation sequencing for forensic SNP analysis. Forensic Sci Int Genet 14:50–60. https://doi.org/10.1016/j.fsigen.2014.08.013
21. Lou C, Cong B, Li S, Fu L, Zhang X, Feng T, Su S, Ma C, Yu F, Ye J, Pei L (2011) A SNaPshot assay for genotyping 44 individual identification single nucleotide polymorphisms. Electrophoresis 32(3–4):368–378. https://doi.org/10.1002/elps.201000426
22. Di Cristofaro J, Silvy M, Chiaroni J, Bailly P (2010) Single PCR multiplex SNaPshot reaction for detection of eleven blood group nucleotide polymorphisms: optimization, validation, and one year of routine clinical use. J Mol Diagn 12(4):453–460. https://doi.org/10.2353/jmoldx.2010.090222
23. Arpan A, Salil V (2016) Single base primer extension assay (SNaPshot) for rapid detection of human immunodeficiency virus-1 drug resistance mutations. J Mol Biomarkers Diagn 7(2):1000271. https://doi.org/10.4172/2155-9929.1000271
24. Tamminen M, Spaak J, Caduff L, Schiff H, Lang R, Schmid S, Montealegre MC, Julian TR (2020) Digital multiplex ligation assay for highly multiplexed screening of beta-lactamase-encoding genes in bacterial isolates. Commun Biol 3(1):264. https://doi.org/10.1038/s42003-020-0980-7
25. Hasib L, Dilcher M, Hufert F, Meyer-Konig U, Weidmann M (2011) Development of a flow-through microarray based reverse transcriptase

multiplex ligation-dependent probe amplification assay for the detection of European Bunyaviruses. Mol Biotechnol 49(2): 176–186. https://doi.org/10.1007/s12033-011-9389-3

26. Zhi J, Hatchwell E (2008) Human MLPA Probe Design (H-MAPD): a probe design tool for both electrophoresis-based and bead-coupled human multiplex ligation-dependent probe amplification assays. BMC Genomics 9: 407. https://doi.org/10.1186/1471-2164-9-407
27. Samelak-Czajka A, Marszalek-Zenczak M, Marcinkowska-Swojak M, Kozlowski P, Figlerowicz M, Zmienko A (2017) MLPA-based analysis of copy number variation in plant populations. Front Plant Sci 8:222. https://doi.org/10.3389/fpls.2017.00222
28. Benard-Slagter A, Zondervan I, de Groot K, Ghazavi F, Sarhadi V, Van Vlierberghe P, De Moerloose B, Schwab C, Vettenranta K, Harrison CJ, Knuutila S, Schouten J, Lammens T, Savola S (2017) Digital multiplex ligation-dependent probe amplification for detection of key copy number alterations in T- and B-cell lymphoblastic leukemia. J Mol Diagn 19(5):659–672. https://doi.org/10.1016/j.jmoldx.2017.05.004
29. Barbano R, Pasculli B, Coco M, Fontana A, Copetti M, Rendina M, Valori VM, Graziano P, Maiello E, Fazio VM, Parrella P (2015) Competitive allele-specific TaqMan PCR (Cast-PCR) is a sensitive, specific and fast method for BRAF V600 mutation detection in Melanoma patients. Sci Rep 5:18592. https://doi.org/10.1038/srep18592
30. Olivier M (2005) The Invader assay for SNP genotyping. Mutat Res 573(1–2):103–110. https://doi.org/10.1016/j.mrfmmm.2004.08.016
31. Kalendar R (2022) A guide to using FASTPCR software for PCR, in silico PCR, and oligonucleotide analysis. Methods Mol Biol 2392: 223–243. https://doi.org/10.1007/978-1-0716-1799-1_16
32. Kalendar R, Khassenov B, Ramankulov Y, Samuilova O, Ivanov KI (2017) FastPCR: an in silico tool for fast primer and probe design and advanced sequence analysis. Genomics 109(3–4):312–319. https://doi.org/10.1016/j.ygeno.2017.05.005
33. Kalendar R, Lee D, Schulman AH (2011) Java web tools for PCR, in silico PCR, and oligonucleotide assembly and analysis. Genomics 98(2):137–144. https://doi.org/10.1016/j.ygeno.2011.04.009

[illegible]

Chapter 2

Digital PCR-Based Genotyping: A Precision Approach to HCMV Drug Resistance

Mojca Milavec, Tašja Cvelbar, and Alexandra Bogožalec Košir

Abstract

The genotyping workflow described uses digital PCR (dPCR) to detect and quantify drug resistance mutations in human cytomegalovirus (HCMV). The method focuses on the detection and quantification of three common mutations in the UL97 gene at codons 460, 594, and 595, which are responsible for the majority of ganciclovir-resistant clinical isolates. The dPCR approach offers high sensitivity and accuracy, making it suitable for routine testing as well as a reference measurement procedure for external quality assessment schemes. The workflow includes several key steps: DNA isolation, preparation of the dPCR reaction mixture, partitioning, thermocycling, and data analysis. This method improves the detection capabilities of HCMV drug resistance and provides a robust and efficient tool for clinical and research applications.

Key words Digital PCR, Human cytomegalovirus, Antimicrobial drug resistance, Mutations

1 Introduction

Human cytomegalovirus (HCMV), or human herpesvirus 5 (HHV-5), is a member of the *Betaherpesvirinae* subfamily and, like other herpesviruses, can establish lifelong latent infections in humans. In healthy, immunocompetent individuals, HCMV is controlled by a vigorous immune response resulting in an asymptomatic infection or causing mild symptoms, while in individuals with suppressed, compromised, or immature immune systems, such as transplant recipients, patients with acquired immunodeficiency disease syndrome, or new-born babies, severe HCMV illness can occur [1–3]. Despite enormous efforts to develop vaccines, there is no licensed HCMV vaccine, yet. The gold standard treatment for preventing HCMV diseases is the antiviral drug ganciclovir. However, a prolonged antiviral therapy, suboptimal drug levels, high levels of immunosuppression, and lack of HCMV immunity can cause development of antiviral resistance due to mutations in UL97

Anthony Torres (ed.), *Genotyping: Methods and Protocols*, Methods in Molecular Biology, vol. 2943,
https://doi.org/10.1007/978-1-0716-4642-7_2,

gene [4–6]. UL97 gene codes for the viral protein kinase, which phosphorylates ganciclovir into its active form of ganciclovir monophosphate, thus mutations in this gene prevent phosphorylation of the pro-drug, providing drug resistance to viral strains.

For routine genotyping, Sanger sequencing is used for detection of viral mutations that confer drug resistance, and it is expected that next-generation sequencing will prevail due to its higher capacity [5]. Another approach, genotyping by dPCR, can provide a rapid, sensitive, and accurate method for detection and quantification of the most common resistance-conferring mutations on the one hand and support the standardization of next-generation sequencing approaches for genotyping on the other.

Here we describe three dPCR methods used for detection and quantification of common mutations at codons 460, 594, and 595 of UL97, which account for 70% of ganciclovir-resistant clinical isolates [7, 8], as well as the reference measurement procedure for HCMV quantification [9, 10]. The analytical method described below is intended for the accurate quantification of control or reference materials or external quality assessment schemes. For this purpose, two or more vials of one batch of the sample are used. If the method is to be used for routine tests, the number of vials and technical parallels can be reduced, and sample dilution step can be omitted.

2 Materials

2.1 DNA Isolation

1. Biosafety cabinet level 2.
2. Water bath/heating block.
3. Vortex.
4. Centrifuge, max speed at least 13,000 × *g*.
5. Pipettes with corresponding tips.
6. High Pure Viral Nucleic Acid Kit (Roche) (*see* **Note 1**).
7. Absolute ethanol (*see* **Note 2**).
8. Nuclease-free water (*see* **Note 3**).
9. Nuclease-free 1.5 mL microcentrifuge tube.
10. DNA low bind 1.5 mL microcentrifuge tubes (*see* **Note 4**).

2.2 dPCR Reaction Setup

1. UV chamber.
2. Pipettes with corresponding tips.
3. Vortex and microcentrifuge.
4. ddPCR™ Supermix for probes (No dUTP) (Bio-Rad).
5. Nuclease-free water.
6. Oligonucleotides (Table 1).

Table 1
Primers and probes used in genotyping for antimicrobial drug resistance in HCMV

Target	Types of oligonucleotide	5′-sequence-3′	Final concentration in PCR (nM)	Amplicon length (bp)
UL54 [12]	UL54 forward primer	5′-GGCCTCGTAGTGAAAAT TAATGGT-3′	600	72
	UL54 reverse primer	5′-GGCCGTTACTGTCTGCAGGA-3′	600	
	UL54 probe	5′-FAM-CCGTATTGGTGCGC GATCTGTTCAA-BHQ-3′	200	
A594V (adapted from [13])	A594V forward primer	5′-ACGGAGGCGTTGCTCTTTAA-3′	900	75
	A594V reverse primer	5′-GGAGCAGTGCGTGAGCTTG-3′	900	
	A594V probe	5′-FAM-CTCCAACACGCGGC-MGBNFQ-3′	300	
L595S (adapted from [13])	L595S forward primer	5′-ACGGAGGCGTTGCTCTTTAA-3′	900	75
	L595S reverse primer	5′-GGAGCAGTGCGTGAGCTTG-3′	900	
	L595S probe	5′-FAM-TTCTCCGACGCGCGG-MGBNFQ-3′	300	
M460V [8]	M460V forward primer	5′-TGTTTCATCACGACCAGTGGAA-3′	900	163
	M460V reverse primer	5′-CTGGGGTTGTACGGGTTCAC-3′	900	
	M460V probe	5′-FAM-ACGTTCACGGGTGTAA-MGBNFQ-3′	300	

2.3 Generation of Partitions and Thermal Cycling

1. UV chamber.
2. Pipettes (Rainin) with corresponding tips (*see* **Note 5**).
3. Holders for cartridges (Bio-Rad).
4. DG8 cartridges (Bio-Rad).
5. DG8 gaskets (Bio-Rad).
6. Droplet generator oil (Bio-Rad).
7. 96-well semi-skirted microtiter plate (e.g., Eppendorf).
8. Cooling block.
9. Pierceable foil heat seal (Bio-Rad).
10. Heat sealer (e.g., Bio-Rad PX1 PCR Plate Sealer).
11. Thermal cycler (e.g., C1000 or T100 Touch Thermal, Bio-Rad).

2.4 Partition Read Out and Analysis

1. Droplet reader oil (Bio-Rad).
2. QX200 Droplet Reader (Bio-Rad).
3. QuantaSoft software (Bio-Rad).
4. Excel (Microsoft) or similar software.

3 Methods

3.1 DNA Isolation

1. Work in the biosafety cabinet in a biosafety level 2 laboratory.
2. Extract each sample in two parallels. Use 200 μL of undiluted primary sample (cell culture lysate, whole blood, serum, and plasma) for each isolation.
3. Included one negative control of isolation (NCI) in each series to demonstrate the absence of sample or kit contamination during isolation (*see* **Note 6**).
4. Follow the isolation protocol High Pure Viral Nucleic Acid Kit manual [11] (*see* **Notes 1–3**).
5. Use isolated DNA directly or store it at −20 °C.

3.2 Analysis Setup

1. QX200 platform enables a maximum of 96 reactions in one rune (96 well plate). An example of a plate setup in Excel for analysis of two samples (A and B), each with two vials, is shown in Fig. 1.
2. Calculate the volume of PCR reactions needed for analysis of all samples. Each final PCR reaction should consist of 10 μL of ddPCR Supermix for probes (No dUTP), 6 μL of primers and probes in appropriate working concentration to achieve final concentrations as described in Table 1 (*see* **Notes 7** and **8**), and 4 μL of DNA. Due to the loss of volume in the pipetting steps,

Target

	UL54		A594V		L595S		M460V			
	1	2	3	4	5	6	7	8	9	10	11	12

	1–2	3–4	5–6	7–8	9	10	11	12
A	A1-1 10x	A1-1 10x	A1-1 10x	A1-1 10x	NCI	PC		
B	A1-2 10x	A1-2 10x	A1-2 10x	A1-2 10x	NCI	PC		
C	A2-1 10x	A2-1 10x	A2-1 10x	A2-1 10x	PC	NTC		
D	A2-2 10x	A2-2 10x	A2-2 10x	A2-2 10x	NTC	NTC		
E	B1-1 10x	B1-1 10x	B1-1 10x	B1-1 10x	PC	PC		
F	B1-2 10x	B1-2 10x	B1-2 10x	B1-2 10x	PC	PC		
G	B2-1 10x	B2-1 10x	B2-1 10x	B2-1 10x	NTC	NTC		
H	B2-2 10x	B2-2 10x	B2-2 10x	B2-2 10x	NTC	NTC		

Fig. 1 An example of a plate setup for two EQA samples (A and B). Each sample is analyzed in two vials (A1, A2, B1, B2), each vial in two extraction parallels (e.g., A1-1 and A1-2), each extraction parallel in two technical repeats prepared in Excel (*10×* 10 times diluted sample, *PC* positive control, *NCI* negative control of isolation, *NTC* no template control)

a 20% surplus should be prepared when calculating volumes. Consequently, if, e.g., 10 reactions are needed for analysis, the quantity of reagents needed should be calculated for 12 reactions (120 μL of ddPCR Supermix for probes (No dUTP) and 72 μL of PPS mix should be used, instead of 100 μL and 60 μL, respectively).

3.3 PCR Reaction Mixture Preparation

1. The PCR reaction is prepared in a dedicated UV cabinet (*see* **Note 9**).
2. Thaw all reaction components such as primers, probe, or a primer and probe mix, and ddPCR Supermix for probes (No dUTP) (*see* **Note 10**) completely at room temperature and mix thoroughly before use. Exposure to light should be minimized.
3. Mix the adequate quantity of dedicated ddPCR Supermix and primers and probes, or primer and probe mixes (*see* **Note 11**) and transfer the mixes with PCR reaction mixtures to a UV cabinet dedicated for addition of nucleic acids.

Prepare an appropriate (e.g., 10-time) dilution of extracted DNA and add 4.4 μL of DNA, positive controls, NCI or no template controls (NTCs) to 17.6 μL of pre-reaction mixture to prepare a 10% surplus of reaction mixture (*see* **Note 12**). Mix and centrifuge.

3.4 Generation of Partitions and Thermal Cycling

1. Turn on the droplet (partition) generator by connecting it to a power source.
2. Prepare holder(s) for cartridges, DG8 cartridges and DG8 gaskets.
3. Insert the cartridge into the cartridge holder.
4. Transfer 20 μL of the PCR reaction mixture to sample wells (middle row) of the cartridge according to plan. Only 20 μL aerosol-barrier (filtered) Rainin pipette tips should be used in this step (*see* **Note 13**).
5. Transfer 70 μL of droplet generation oil into each oil well (bottom row) of the cartridge.
6. Using the holes on the sides, hook the gasket over the cartridge holder.
7. Open QX200 Droplet Generator by pressing the button on the top of the instrument.
8. Place the holder with the cartridge into the instrument and check if cartridge status indicator light (middle) turned green.
9. Close the instrument by pressing the button on the top of the instrument. When the lid is closed, partition generation begins, indicated by flashing green light on the right. Partition generation is finished when all three indicator lights are solid green.

Table 2
PCR program for amplification of antimicrobial drug resistance and HCMV specific targets

No. of cycles	Time	Temperature (°C)
1	10 min	95
45	15 s 60 s	95 60
1	10 min	98
1	Infinite	4

Lid temperature: 105 °C
Volume: 40 μL
Ramp rate: 2.5 °C/s

10. Open the instrument by pressing the button on the top and remove the holder with the cartridge.
11. Place the holder with the cartridge on a flat surface.
12. Remove and discard the gasket.
13. Transfer 40 μL of suspension with droplets (partitions) from droplets well (top) into a 96-well plate placed on a cooling block according to the plan (*see* **Note 14**).
14. Repeat droplet generation and transfer steps for all the samples.
15. After the transfer of all the samples to a 96-well plate, seal the plate with pierceable foil using PCR plate sealer (*see* **Note 15**).
16. Place sealed 96-well plate in the PCR cycler. Run the program described in Table 2.

3.5 Partition Read Out and Analysis

1. Turn on the droplet reader to warm up the instrument.
2. Start the QuantaSoft software on the connected computer.
3. Transfer the 96-well plate into the QX200 Droplet Reader.
4. Flush the instrument by selecting "Setup" → "Instrument routines" → "Flush System" in the QuantaSoft software.
5. To configure the plate layout select "Setup" → "Template" → "New." Enter the new file name. Enter all the necessary information (sample names, targets, detection type, supermix) using well editor. Save the created template.
6. Click "Run" to initiate droplet reading. Set dye (FAM/HEX in our example) when the Run options window appears.
7. After reading is finished, save the experiment data, and remove the plate from the instrument.

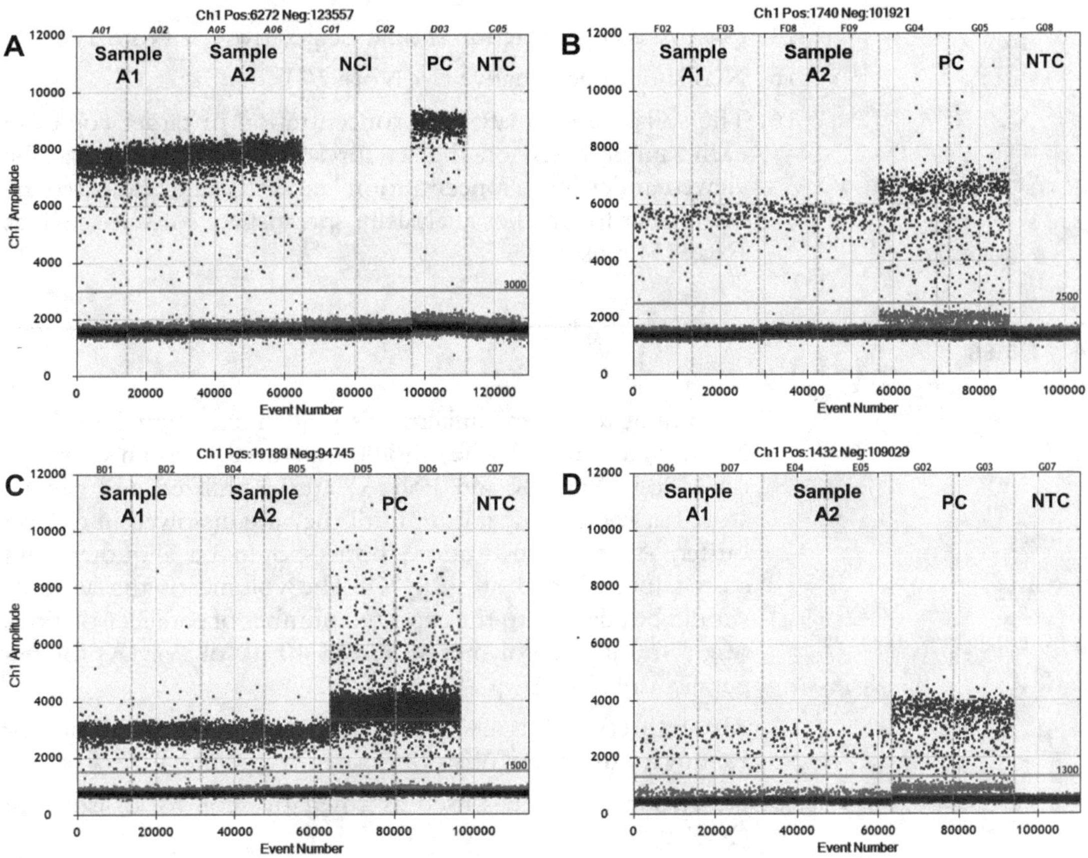

Fig. 2 An example of amplification plots showing both positive and negative results for (**A**) UL54 method; (**B**) A594V mutation detection method; (**C**) L595S mutation detection method; (**D**) M460V mutation detection method (*NCI* negative control of isolation, *NTC* no template control, *PC* positive control—synthetic DNA comprising all targeted sequences)

8. In the QuantaSoft software, select "Setup" → "Plate" → "Load." When the plate is loaded, click "Analyze" to start data analysis.
9. In "1D Amplitude" view, set the fluorescence thresholds for distinguishing between positive and negative droplets. The thresholds should be set simultaneously for all samples in a given assay. NTCs can be used as a guide for determining the negative population and threshold setting just above the negative cluster found in NTC(s) (*see* **Note 16**) (Fig. 2).
10. Before further analysis, quality control should be performed:
11. Wells with accepted number of partitions <10,000 should not be considered for further analysis.
12. The reaction in each well is considered as positive, if three or more partitions show the fluorescent signal above the threshold value.
13. NTC should be negative (*see* **Note 17**).

14. Positive control reaction should be positive (*see* **Note 17**).
15. NCI should be negative (*see* **Note 17**).
16. The software calculates the concentration of target copies in each sample based on the set thresholds. The results can be viewed under the "Concentration" tab and can be exported in . csv format for further analysis in spreadsheet software such as Excel (Microsoft).

4 Notes

1. According to the manufacturer's guidelines, High Pure Viral Nucleic Acid Kit (Roche) should be stored at room temperature. Polymerase K and Poly(A) are lyophilized and can be stored at room temperature until they are dissolved in elution buffer. After reconstitution, both should be aliquoted and stored in dark and at −20 °C. The volume of the aliquots should be adapted to the expected number of parallel isolations (e.g., 550 μL of Polymerase K and 40 μL of Poly(A) for ten parallel isolations).
2. Absolute ethanol is needed for the preparation of Inhibitor Removal Buffer and Wash Buffer.
3. Nuclease-free water is used as a negative control of isolation (NCI).
4. DNA low bind (e.g., Eppendorf DNA LoBind tubes) microcentrifuge tubes are used to maximize recovery of DNA with minimal sample-to-surface binding.
5. According to recommendations from dPCR platform manufacturer, Rainin L-20 XLS 2–20 μL pipette and GP-L10F tips (Rainin 17002429; Tips LTS 20 μL Filter) are to be used for reaction mixture pipetting, and Rainin L-50 XLS 5–50 μL pipette and GP-L200F tips (Rainin 17002428; Tips LTS 200 μL Filter) for droplet pipetting.
6. Negative controls of isolation comprise the same volume of nuclease-free water as is the volume of the sample, e.g., if the sample volume is 200 μL, then 200 μL of nuclease-free water will be needed as a NCI. NCI should be added in to each extraction as the last sample in the series. If there are ten or more samples in the extraction series, an additional NCI should be added. For example, for 12 samples, add 2 NCI, and for 24 samples, add 3, and so on.
7. If oligonucleotides are delivered lyophilized, they need to be resuspended according to manufacturer's instruction to the concentration of 100 μM using nuclease-free water. Lyophilized stock is initially shortly centrifuged to allow all of the

content to sediment. Then it is resuspended with nuclease-free water, well vortexed and centrifuged again. Exposure to light should be minimized. Oligonucleotides are stable up to 6 years when stored at −20 °C. Potential exception are probes with less stable fluorophores.

8. From stock solutions (100 μM) of oligonucleotides, working solutions of forward and reverse primers and probe can be prepared in a form of a primer and probe mix to decrease run-to-run variability and shorten the time for PCR setup. Below is an example of preparation of such mixes for oligonucleotides listed in Table 1 (calculations based on 6 μL mix added to final volume of 20 μL per reaction):
   (a) UL54: UL54 forward primer 36 μL, UL54 reverse primer 36 μL, UL54 probe 12 μL, nuclease-free water 1716 μL.
   (b) A594V: A594V forward primer 54 μL, A594V reverse primer 54 μL, A594V probe 18 μL, nuclease-free water 1674 μL.
   (c) L595S: L595S forward primer 54 μL, L595S reverse primer 54 μL, L595S probe 18 μL, nuclease-free water 1674 μL.
   (d) M460V: M460V forward primer 54 μL, M460V reverse primer 54 μL, M460V probe 18 μL, nuclease-free water 1674 μL.

   Such mixes are stored at −20 °C and are stable for up to 6 years.

9. In order to prevent cross-contamination, it is recommended to perform preparation of PCR reaction mixture and addition of DNA at separate locations using dedicated laboratory equipment.
10. ddPCR Supermix for probes (No dUTP) is a two times concentrated reaction mixture stable at −20 °C till the expiration date. Once thawed, it can be stored at 4 °C for up to 2 weeks.
11. For example, 240 μL of ddPCR Supermix and 132 μL of primers and probe mix should be prepared for each target.
12. To ensure that 20 μL of reaction mixture are transferred to the droplet generation process, the reaction mixture is prepared with a 10% surplus. Diluted DNA should be used shortly after the preparation.
13. For best droplet generation, air bubbles should be avoided and samples should wet the bottom of the wells, so they are wicked into microchannels. To achieve this, pipetting techniques described in the manufacturer's instruction manual (Catalog Number 1864002, Bio-Rad) should be followed. In short, pipette tip should be carefully slid along the wall at a 15°

angle until it passing over the ridge near the bottom. Then the pipette tip should be drag against the bottom of the sample well while a small portion of the sample should be slowly dispensed. After dispensing about half of the sample, the tip should be slowly pull up the side wall while the rest of the sample should be dispensed.

14. To avoid evaporation, place the 96-well plate on a cooling block. To avoid shearing or coalescing the droplets (partitions) pipetting techniques described in the manufacturer's instruction manual (Catalog Number 1864002, Bio-Rad) should be followed. In short, an 8-channel manual L-50 pipette with 200 μL tips should be used. Pipette tips should be positioned at a 30–45° angle to the junction where the side wall meets the bottom of the droplet wells. Suspension with the droplets (partitions) should be slowly drawn from the wells into the pipette tips. Pipette tips should be then positioned along the side of the wells of 96-well plate, near the bottom of the well, and suspension with the droplets should be slowly dispensed. Drawing and aspirating of droplets should take 5 s each. To prevent contamination, full wells should be covered with 8-cap strips.
15. It is recommended to begin thermal cycling within 30 min of sealing the plate. Alternatively, sealed plates can be stored at 4 ° C for up to 4 h before thermal cycling.
16. In our setup, thresholds are set for each method (at amplitude approximately 3300, 3000, and 1300 for the methods targeting A594 V, L595S, and M460 V, respectively, and for the UL54 method at amplitude approximately 3000) in "1D Amplitude" view. As some cross reactivity was observed with wild type (*see* PC in Fig. 2), thresholds were set above the fog observed in PC.
17. The analysis is not valid, if the NTC(s) and/or NCI are positive or if the positive control reaction is negative. It is sufficient to test NCI only for the target that is expected in highest concentration (UL54 in our case).

Conflict of Interest

The authors have no conflicts of interest to declare that are relevant to the content of this chapter.

References

1. Chevillotte M, von Einem J, Meier BM et al (2010) A new tool linking human cytomegalovirus drug resistance mutations to resistance phenotypes. Antivir Res 85:318–327. https://doi.org/10.1016/j.antiviral.2009.10.004
2. Seitz R (2010) Human cytomegalovirus (HCMV)—revised. Transfus Med Hemotherapy 37:365–375. https://doi.org/10.1159/000322141
3. Griffiths P, Reeves M (2021) Pathogenesis of human cytomegalovirus in the immunocompromised host. Nat Rev Microbiol 19:759–773. https://doi.org/10.1038/s41579-021-00582-z
4. Krishna BA, Wills MR, Sinclair JH (2019) Advances in the treatment of cytomegalovirus. Br Med Bull 131:5–17. https://doi.org/10.1093/bmb/ldz031
5. Chou S (2020) Advances in the genotypic diagnosis of cytomegalovirus antiviral drug resistance. Antivir Res 176:104711. https://doi.org/10.1016/j.antiviral.2020.104711
6. Gilbert C, Boivin G (2005) Human cytomegalovirus resistance to antiviral drugs. Antimicrob Agents Chemother 49:873–883. https://doi.org/10.1128/AAC.49.3.873-883.2005
7. Boivin G, Gilbert C, Gaudreau A et al (2001) Rate of emergence of cytomegalovirus (CMV) mutations in leukocytes of patients with acquired immunodeficiency syndrome who are receiving Valganciclovir as induction and maintenance therapy for CMV retinitis. J Infect Dis 184:1598–1602. https://doi.org/10.1086/324672
8. Bogožalec Košir A, Cvelbar T, Kammel M et al (2020) Digital PCR method for detection and quantification of specific antimicrobial drug-resistance mutations in human cytomegalovirus. J Virol Methods 281:113864. https://doi.org/10.1016/j.jviromet.2020.113864
9. Pavšič J, Devonshire A, Blejec A et al (2017) Inter-laboratory assessment of different digital PCR platforms for quantification of human cytomegalovirus DNA. Anal Bioanal Chem 409:2601–2614. https://doi.org/10.1007/s00216-017-0206-0
10. Milavec M, Pavšič J, Bogožalec Košir A et al (2021) The performance of human cytomegalovirus digital PCR reference measurement procedure in seven external quality assessment schemes over four years. Methods 201:65. https://doi.org/10.1016/j.ymeth.2021.03.016
11. Roche Diagnostics GmbH (2020) High pure viral nucleic acid kit. Roche Diagnostics GmbH 20:3–11
12. Sassenscheidt J, Rohayem J, Illmer T, Bandt D (2006) Detection of β-herpesviruses in allogenic stem cell recipients by quantitative real-time PCR. J Virol Methods 138:40–48. https://doi.org/10.1016/j.jviromet.2006.07.015
13. Volfova P, Lengerova M, Lochmanova J et al (2014) Detecting human cytomegalovirus drug resistant mutations and monitoring the emergence of resistant strains using real-time PCR. J Clin Virol 61:270–274. https://doi.org/10.1016/j.jcv.2014.07.008

References

Chapter 3

Digital PCR Genotyping of Pepino Mosaic Virus

Amadej Jelenčič, David Dobnik, Alexandra Bogožalec Košir, and Nataša Mehle

Abstract

Pepino mosaic virus (PepMV) is a plant pathogen causing significant economic losses in tomato production. Sensitive, reliable, and robust detection methods are crucial for containing the spread of PepMV and reducing its damaging effects. Digital PCR (dPCR) presents several advantages to conventional real-time quantitative PCR (qPCR), including absolute quantification ability, robust quantitative multiplexing capabilities, and straightforward result analysis. Furthermore, dPCR is especially suitable for analysis of complex samples due to its remarkable tolerance to PCR inhibitors, which makes it a promising method for plant virus genotyping. In this chapter, we present two protocols for PepMV genotyping and quantification using one-step reverse transcription digital PCR (RT-dPCR). The first protocol outlines four simplex assays, while the second describes two duplex assays for precise and comprehensive genotyping of PepMV variants.

Key words Plant virus, Detection, Quantification, Pepino mosaic virus, PepMV, Digital PCR, dPCR, RT-dPCR, Duplex

1 Introduction

Pepino mosaic virus (PepMV) is a non-enveloped rod shaped, single-stranded RNA virus belonging to the genus *Potexvirus* [1]. It is a pathogen that affects the genus *Solanum*, and its effects on greenhouse tomato production are economically very damaging [2–4]. PepMV exhibits significant genomic diversity, and five main genotypes are currently recognized: Peruvian (LP); European (Eur); American (US1, US2); and Chilean (Ch2) [5, 6]. Highly virulent and damaging isolates are distinguished from mild isolates by single nucleotide variations [7]. The enforcement of strict hygiene measures required to control the spread of PepMV is challenged by the virus's remarkable resilience [8]. Cross protection strategies have been researched that entail the infection of a plant with a mild strain of a virus to confer resistance to more severe strains [9]. One such strategy, the PMV-1 vaccine has been

Anthony Torres (ed.), *Genotyping: Methods and Protocols*, Methods in Molecular Biology, vol. 2943, https://doi.org/10.1007/978-1-0716-4642-7_3,

approved in the European Union [10]. There has also been research in controlling the spread of PepMV by RNA interference using transgenesis [11]. Reliable detection, quantification, and genotyping of PepMV are essential not only for developing, applying, and evaluating approaches to combat it, but also for assessing plant resistance to PepMV, determining PepMV concentrations across various plant tissues and organs, and measuring the prevalence of different PepMV strains in mixed infections. As PepMV can be transmitted through seeds [3, 12], quantification and genotyping are also important for screening of seed stocks to prevent major economic damage [13].

Reverse transcription real-time quantitative polymerase chain reaction (RT-qPCR) is a commonly employed method for PepMV quantification and genotyping [13, 14]. This method only allows for relative quantification, as absolute quantification would require standard curves prepared from certified reference materials, which are currently not available for PepMV. The method is also heavily affected by PCR inhibition caused by various substances commonly present in plant material [15].

Conversely, digital PCR (dPCR) provides absolute quantification of target sequences without requiring standard curves [16]. It has also shown remarkable tolerance to PCR inhibitors, improving the accuracy and repeatability of the results [17, 18]. Further advantages of the method include its robust multiplexing (detection of more than one target in one reaction) capabilities and straightforward analysis of results [19–22].

Here we describe two one-step RT-dPCR-based protocols for accurate genotyping and quantification of PepMV. The first protocol was developed for the Bio-Rad QX200 system (suitable also for QX One and QX600) and consists of the following simplex (one target) assays for detection and quantification of PepMV genotypes: PepMV-universal for all genotypes, PepMV-Eur for Peruvian and European genotypes, PepMV-Ch2 for the Ch2 and US2 genotypes, and PepMV-US1 for the US1 genotype. The probes were labeled with 6-carboxyfluorescein (FAM). The second protocol was developed for the Stilla Naica platform and describes two duplex assays acquired by combining the simplex assays. The first duplex consists of the PepMV-US1 and PepMV-Universal assays, while the second duplex contains the primers and probes for the PepMV-Eur and PepMV-Ch2 assays. Here the dyes on two probes differ from the ones used in the first protocol, namely hexachloro-fluorescein (HEX) and cyanine-5 (Cy5) are used instead of FAM for the PepMV-Ch2 and PepMV-US1 assays, respectively. Generally, the level of background fluorescence is higher in multiplexed assays due to larger quantities of fluorescent dyes present. Therefore, double-quenched ZEN probes with non-fluorescent quencher Iowa Black FQ (IBFQ) on 3′ end are used in the duplex assays to reduce the background fluorescence.

2 Materials

2.1 Samples

1. Extracted sample DNA (*see* **Note 1**).
2. Positive and negative controls (*see* **Note 2**).

2.2 Simplex Assays on the Bio-Rad QX200 Platform

2.2.1 Reaction Preparation and Thermal Cycling

1. One-Step RT-ddPCR Advanced Kit for Probes (Bio-Rad, USA; includes the master mix, reverse transcriptase, and dithiothreitol (DTT)).
2. Rainin pipettes with corresponding presterilized BioClean LTS tips with filter (*see* **Note 3**).
3. Two UV chambers (*see* **Note 4**).
4. Pierceable foil heat seal (Bio-Rad, USA) or equivalent.
5. Heat sealer (Eppendorf, Germany) or equivalent.
6. T100™ Thermal Cycler (Bio-Rad, USA) or equivalent.
7. Pipettes.
8. Nuclease-free microcentrifuge tubes.
9. twin.tec semi-skirted 96-well plates (Eppendorf, Germany).
10. Vortex and microcentrifuge.
11. Centrifuge with rotor for 96-well plate centrifugation.

2.2.2 Droplet Generation and Readout

1. QX200™ Droplet Generator or Automated droplet generator (AutoDG) (Bio-Rad, USA).
2. DG8 Cartridge Holders, Cartridges, and Gaskets (for QX200 droplet generator) (Bio-Rad, USA).
3. Droplet Generation Oil for Probes (for QX200 droplet generator) (Bio-Rad, USA).
4. DG32 Automated Droplet Generator Cartridges (for AutoDG) (Bio-Rad, USA).
5. Automated droplet Generation Oil for Probes (for AutoDG) (Bio-Rad, USA).
6. Pipette Tips for the AutoDG™ System (Bio-Rad, USA).
7. Pipette Tip Waste Bins for the AutoDG™ System (Bio-Rad, USA).
8. Nuclease-free water.
9. Droplet Reader Oil (Bio-Rad, USA).
10. QX200™ Droplet Reader (Bio-Rad, USA).
11. QuantaSoft analysis software or QXManager Software (Bio-Rad, USA).
12. Cooling blocks.

2.3 Duplex Assays on the Stilla Naica Platform

2.3.1 Reaction Preparation

1. qScript XLT One-Step RT-qPCR ToughMix (no ROX) (Quanta BioSciences, USA) or equivalent (*see* **Note 5**).
2. Fluorescein (FITC; 1 μM).
3. Pipettes.
4. Nuclease-free microcentrifuge tubes.
5. Vortex and microcentrifuge.

2.3.2 Digital PCR and Data Analysis

1. Geode partitioning and thermocycling instrument (Stilla Technologies, France).
2. Prism3 or Prism6 detection and imaging instrument (Stilla Technologies, France).
3. Sapphire chips (Stilla Technologies, France).
4. Antistatic wipes.
5. Low-lint, high absorbency, and chemically inert wipes.
6. Crystal Reader and Crystal Miner software (Stilla Technologies, France).

3 Methods

3.1 Simplex Assays on the Bio-Rad QX200 Platform

3.1.1 Preparation of Reaction Mixtures

1. Prepare the reaction mixtures by mixing in nuclease-free tubes of 5 μL master mix with 1 μL DTT (300 mM), 2 μL reverse transcriptase, nuclease-free water, primers, and probes, according to Table 1, up to a final volume of 16 μL (*see* **Note 5**). Vortex the mixtures and centrifuge them briefly (*see* **Note 6**).
2. Distribute the mixtures into 96-well polypropylene plates (*see* **Note 7**).

Table 1
Primers and probes used in simplex PepMV genotyping assays

Target and method	Name	DNA sequence of the oligonucleotide (5′-sequence-3′)	Final concentration in the PCR (nmol/L)
PepMV-Universal	KL05-48 forward 1	ACTCCTAGAGCTGACCTCAC	200
	KL05-51 reverse 1	TCTCCAGCAACAGGTTGGTA	200
	KL05-49 forward 2	ACTCCTAGAGCTGATCTTAC	200
	KL05-52 reverse 2	TCACCTGCAACTGGTTGATA	200
	KL05-50 probe	FAM-TGTCAGCTTGCATTTACTTCCAAAA-TAMRA	400
PepMV-Eur	Eur-rep forward	TGGGTCAAGAAAGTGGAAAAATTAG	900
	Eur-rep reverse	TGCATGAAAGCTGCAATGGT	900
	Eur-rep probe	FAM-TGCTGTCAAGTCAAAGCCTGGCCA-TAMRA	200

(continued)

Table 1
(continued)

Target and method	Name	DNA sequence of the oligonucleotide (5′-sequence-3′)	Final concentration in the PCR (nmol/L)
PepMV-Ch2	Ch2-cp forward	TGGGTTTAGCAGCCAATGAGA	900
	Ch2-cp reverse	AACTTTGCACATCAGCATAAGCA	900
	Ch2-cp probe	FAM-CGGACCTGCCATGTGGGAC CTC-TAMRA	200
PepMV-US1	US1-cp forward	AGCGCGTGCTTATGCTGAT	900
	US1-cp reverse	CGTGAGAGTGCTGGATTTGAAG	900
	US1-cp probe	FAM-CCGCACAACTCATAGGT GCCACCC-TAMRA	200

3. Add 4 μL of each RNA sample or non-template control (NTC) into the wells containing the reaction mixtures (*see* **Note 5**). Seal the 96-well plates with pierceable foil. Vortex and centrifuge briefly.

3.1.2 Manual Droplet Generation Using Bio-Rad QX200™ Droplet Generator

1. Turn on the droplet generator by connecting it to a power source.
2. Insert the DG8 cartridge into the cartridge holder.
3. Transfer 20 μL of each reaction sample into the sample wells on the cartridge.
4. Transfer 70 μL of droplet generation oil into each oil well on the cartridge.
5. Secure the DG8 cartridge gasket over the cartridge.
6. Press the button on the droplet generator to open it.
7. Place the holder with the cartridge into the droplet generator.
8. Close the lid on the droplet generator to begin droplet generation. After completion (indicated by all lights shining green), remove the holder with the cartridge from the instrument and remove the gasket from the cartridge.
9. Transfer the droplet suspensions (40 μL) onto a 96-well plate placed on a cooling block.
10. Repeat the droplet generation and plate transfer steps for all the samples.
11. Seal the plate harboring the droplets with pierceable foil (*see* **Note 8**). Do not vortex or centrifuge the plate.

3.1.3 Automatic Droplet Generation Using Bio-Rad AutoDG

1. Turn on the automatic droplet generator and configure the sample plate in the instrument software.
2. Place the DG32 cartridges, pipette tip boxes without box lids (*see* **Note 9**) and the tip waste bin into their dedicated holders in the instrument (as indicated on the instrument screen).
3. Place the sealed 96-well plate containing the prepared reactions into its dedicated holder (named "sample plate") in the instrument.
4. Place the cooling block (*see* **Note 10**) into the holder named "droplet plate" and place an empty 96-well plate into the cooling block for droplet collection.
5. Check if there is enough automated droplet generation oil present and change the oil if needed (*see* **Note 11**).
6. Tap "START Droplet Generation" and "Start Run" to initiate droplet generation.
7. After the run is finished and the instrument door unlocked, remove the plate with droplets from the instrument and discard the used consumables.
8. Seal the plate harboring the droplets with pierceable foil (*see* **Note 8**). Do not vortex or centrifuge the plate.

3.1.4 Thermal Cycling

1. Transfer the plate into a thermocycler and run the PCR under the conditions shown in Table 2.

3.1.5 Droplet Reading and Data Analysis

1. Transfer the 96-well plate into the QX200 Droplet Reader (*see* **Note 12**).
2. Run the QuantaSoft software on the connected computer.
3. Flush the instrument by clicking "Setup" → "Instrument routines" → "Flush System" in the QuantaSoft software.
4. Click "Setup" → "Template" → "New" to configure the plate layout and enter the necessary information. Save the created template.

Table 2
Thermal cycling conditions for the QX200 platform

Cycling step	*T* (°C)[a]	Time	Number of cycles
Reverse transcription	50	60 min	1
Enzyme activation	95	10 min	1
Denaturation	95	30 s	40
Annealing and elongation	56	1 min	
Enzyme deactivation	98	10 min	1

Ramp rate should be set to 2.5 °C/s

5. Click "Run" to initiate droplet reading.
6. After reading is finished, save the experiment data, and remove the plate from the instrument.
7. Open the experiment in the QuantaSoft software and click "Analyze" to start data analysis.
8. Imaging results are visualized as a scatterplot of fluorescence amplitudes for each droplet (Fig. 1). Set the fluorescence thresholds for distinguishing between positive and negative droplets. The thresholds can be set simultaneously for all

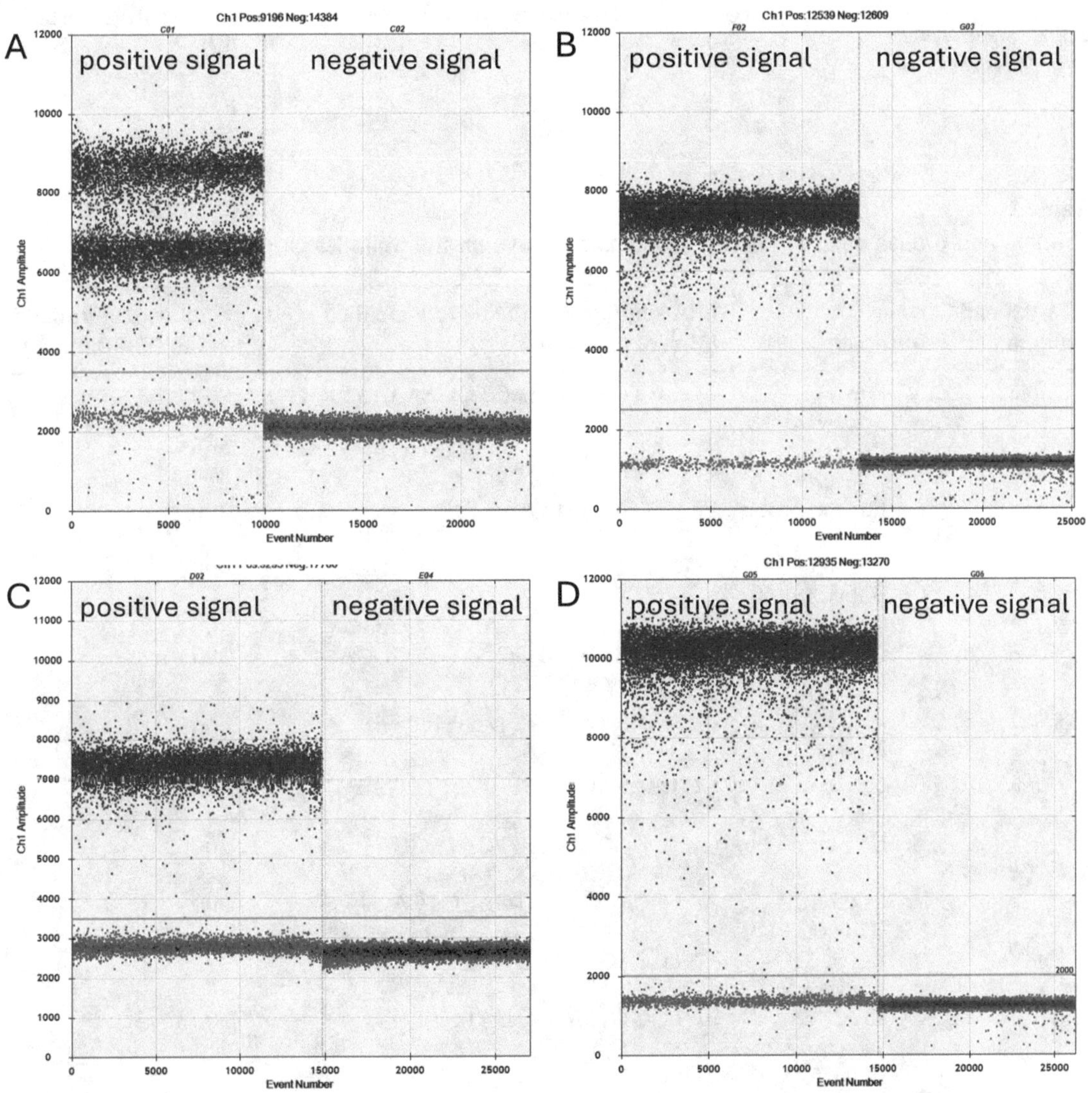

Fig. 1 Amplification plots for simplex assays on QX200. The left part of each panel represents a positive reaction (positive signal), while the right sides of the panel represent non-template controls (negative signal). The pink line represents the set threshold. (**a**) Universal assay, (**b**) EUR assay, (**c**) CH2 assay, (**d**) US2 assay

samples in a given assay. Use NTCs as a guide for determining the negative population and threshold setting (*see* **Note 13**).

9. The software calculates the concentration of target copies in each sample based on the set thresholds. The results can be viewed under the "Concentration" tab and can be exported in .csv format for further analysis in spreadsheet software such as Excel (Microsoft).

3.2 Duplex Assays on the Stilla Naica Platform

3.2.1 Preparation of Reaction Mixtures

1. Prepare the reaction mixtures by mixing 13.75 μL ToughMix (2×) (no ROX) with 2.75 μL FITC (final concentration, 100 nM), nuclease-free water, primers, and probes, according to Table 3 (*see* **Notes 6** and **14**). Vortex and centrifuge briefly.

Table 3
Primers and probes used in PepMV genotyping assays on the Stilla Naica platform

Target and method	Name	DNA sequence of the oligonucleotide (5′-sequence-3′)	Final concentration in the PCR (nmol/L)
Duplex 1 (Eur + Ch2)	Eur-rep forward	TGGGTCAAGAAAGTGGAAAAATTAG	900
	Eur-rep reverse	TGCATGAAAGCTGCAATGGT	900
	Eur-rep probe	FAM-TGCTGTCAAGT CAAAGCCTGGCCA-ZEN/IBFQ	200
	Ch2-cp forward	TGGGTTTAGCAGCCAATGAGA	900
	Ch2-cp reverse	AACTTTGCACATCAGCATAAGCA	900
	Ch2-cp probe	HEX-CGGACCTGCCATGTGGGA CCTC-ZEN/IBFQ	200
Duplex 2 (US1 + universal)	US1-cp forward	AGCGCGTGCTTATGCTGAT	900
	US1-cp reverse	CGTGAGAGTGCTGGATTTGAAG	900
	US1-cp probe	Cy5-CCGCACAACTCATAGGT GCCACCC-ZEN/IBFQ	200
	KL05-48 forward 1	ACTCCTAGAGCTGACCTCAC	200
	KL05-51 reverse 1	TCTCCAGCAACAGGTTGGTA	200
	KL05-49 forward 2	ACTCCTAGAGCTGATCTTAC	200
	KL05-52 reverse 2	TCACCTGCAACTGGTTGATA	200
	KL05-50 probe	FAM-TGTCAGCTTGCATTTACTTC CAAAA-ZEN/IBFQ	400

3.2.2 Droplet Generation and Thermal Cycling

1. Clean the bottom of each chip by sliding it on an antistatic wipe and a low-lint, high absorbency and chemically inert wipe.
2. Remove the white caps off the Naica Ssapphire chip inlets and load 25 μL of each reaction mixture into the inlets. Do not insert the pipette tip into the oil phase. Close the inlets with tall caps included with the chips.
3. Turn on the power switch at the back panel of the Geode.
4. Connect the pressure source to the pressure inlet at the back of the instrument and switch on the power source (*see* **Note 15**).
5. Place the chips into their dedicated spaces in the instrument.
6. Set the partitioning and thermal cycling conditions as indicated in Table 4 (*see* **Note 16**).
7. Press the "Run" icon on the instrument screen, press the "Open" icon, and select the correct cycling program in the "local/templates" subfolder.
8. Press the "Open" icon and then the "Play" icon to start the program.
9. Once the run is over, touch the screen, open the lid of the instrument, remove the Sapphire chips, and close the lid.

3.2.3 Droplet Imaging and Data Analysis

1. Turn on the power switch at the back of the Prism instrument and press the wake-up button on the front panel.
2. Log in with your username and password.
3. Click on the "Crystal Reader" icon on the instrument or a computer to launch the reader software.
4. Click on the "OPEN TRAY" button in the software or press the "Eject" button on the instrument to eject the tray.
5. Insert the Sapphire chips in the Sapphire tray holder.

Table 4
Cycling conditions for the Naica Geode

Cycling step	*T* (°C)	Pressure (mbar)	Time	Number of cycles
Partition	40	Atmospheric pressure to +950	12 min	1
cDNA synthesis	50	+950	10 min	1
Enzyme activation	95	+950	1 min	1
Denaturation	95	+950	30 s	45
Annealing and elongation	56	+950	15 s	
Release	Down to 25	Down to atmospheric pressure	33 min	1

6. Place the tray holder on the tray and gently push it toward the back of the tray.
7. Click on the "CLOSE TRAY" button in the software or press the "Eject" button on the instrument to retract the tray.
8. Click on the "NEW EXPERIMENT" button in the software, press "BROWSE," and select the "ScanningTemplate_-Prism6_SapphireChip_QuantaBio-qScript-XLT-1-step-RT-qPCR-Thoughmix_Taqman_v1.1.ncx" template.
9. Enter the name of the experiment in the "New Experiment" field and add potential additional information into the "Comments" field.
10. Under the "Experiment Details" tab, check the detection channels you want to utilize and enter the target and fluorophore name for each one (*see* **Note 17**).
11. Select the chambers to scan and enter the sample name and type for each chamber.
12. For multiplex assays, a color compensation matrix needs to be prepared according to manufacturer instructions.
13. Click on the "SCAN" button, define the location to save scanned data, and confirm that the correct chips are loaded to initiate the scanning of droplets.
14. After the scanning is completed, click "OK" to display the results.
15. Open the experiment file in the Crystal Miner software.
16. Imaging results are visualized as a scatterplot of fluorescence amplitudes for each droplet (Fig. 2). Set the fluorescence thresholds for determining positive and negative droplets. It is advised to set the thresholds in the 2D view to account for possible crosstalk between the fluorescence channels (Fig. 3).
17. Quantification results are presented by the software as the number of target copies per μL reaction. The results can be exported in .csv format for further analysis in spreadsheet software.

4 Notes

1. Different methods can be used to extract RNA from plant material. For the development of the assays described here, RNeasy Plant mini kit (QIAGEN, Germany) was used with some modifications to the manufacturer's instructions. Briefly, the protocol involved the homogenization of 200 mg of plant material in 900 μL of lysis buffer, followed by loading 600 μL of the homogenate onto RNeasy Mini Spin columns. RNA was eluted from the columns by applying two consecutive aliquots of 50 μL each (i.e., 100 μL total) of RNase-free warm water (65 °C).

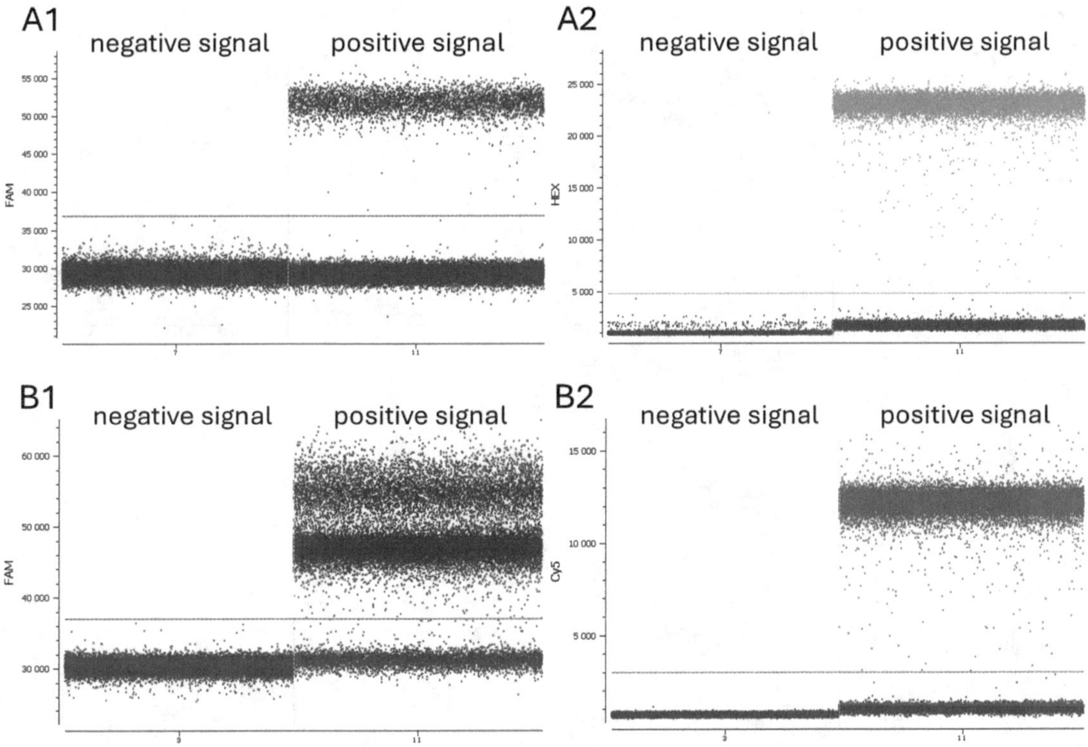

Fig. 2 Amplification plots for duplex assays on Naica. The left part of each panel represents the negative template control (negative signal), while the right sides of the panel represent positive reactions (positive signal). The blue, green, and red lines represent the set thresholds. Panels **a1** and **a2** present the EUR + Ch2 duplex (**a1**—EUR target, **a2**—Ch2+US2 target), and panels **b1** and **b2** present the US1 + universal duplex (**b1**—universal target, **b2**—US1 target)

2. At least non-template controls (NTC) and positive controls containing DNA of tested target(s) should be included. It is recommended to also include a negative extraction control to test for potential contamination during RNA extraction and positive extraction controls, which can be either internal or external. Internal positive controls are prepared for each sample and consist of RNA extracted from samples spiked with exogenous nucleic acid that has no relation with the target virus or endogenous nucleic acid and is used to confirm successful extractions. Naturally infected host tissue or spiked healthy host tissue can serve as an external positive control. During initial assay testing and the validation process, an additional control should be incorporated consisting of healthy material of the host plant to assess the background signal produced by the matrix.
3. Use of Rainin tips is recommended for all steps of the protocol, as well as upstream sample preparation. Other tips can have rough tip openings which can lead to shedding of plastic debris in the reaction mixture that can cause shearing of droplets and detection of artifacts during droplet imaging.

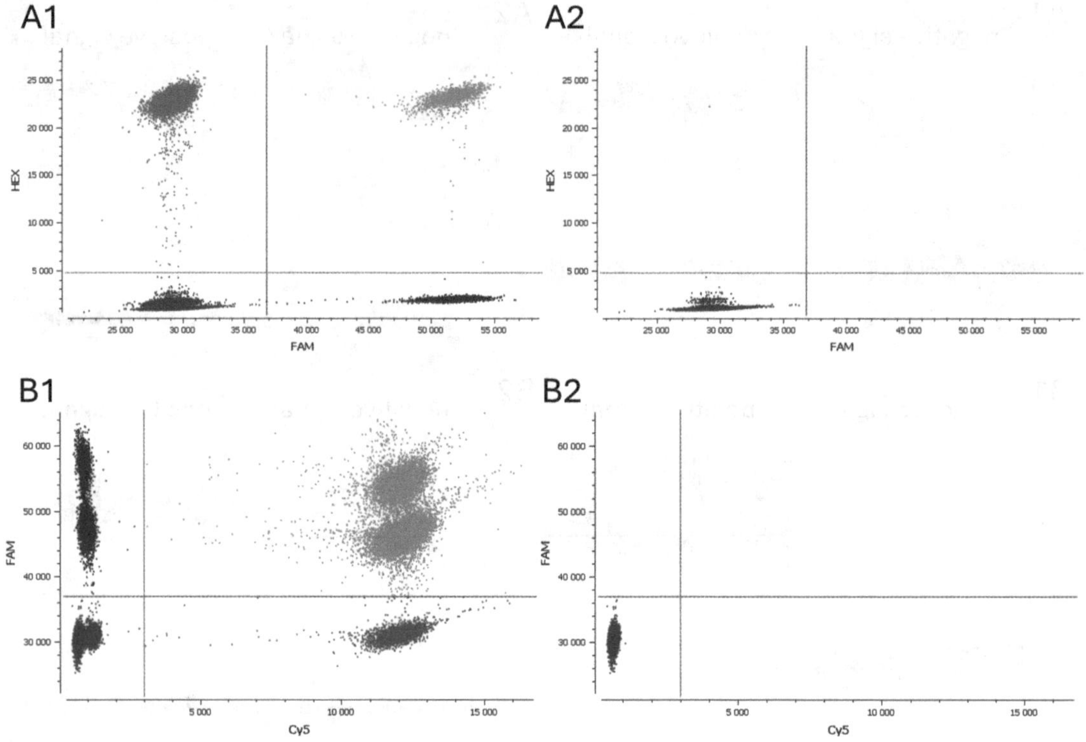

Fig. 3 Amplification plots for duplex assays on Naica in 2D view. The blue, green, and red lines represent the set thresholds. Panels **a1** and **a2** present the EUR + Ch2 duplex (**a1**—positive signal, **a2**—NTC), and panels **b1** and **b2** present the US1 + universal duplex (**b1**—positive signal, **b2**—NTC)

4. This is recommended as the preparation of reaction mixture and addition of samples should be performed at separate locations using dedicated laboratory equipment to prevent cross-contamination. Both steps should be performed at locations different from the one used for RNA extraction.
5. It is recommended to prepare the mixtures with 10% excess volume to offset pipetting losses.
6. It is important to mix the reaction mixtures well, as failure to do so can result in reduced reproducibility of results due to the viscosity of the master mix. Vortex with lower speeds to prevent bubble formation.
7. Use of polypropylene plates is recommended to minimize DNA binding to the plate and thus improve repeatability. Plates must be semi-skirted to fit into the instrument. Eppendorf twin.tec semi-skirted plates and Bio-Rad ddPCR 96-well plates have been proven to be suitable for this application.

8. It is recommended to begin thermal cycling within 30 min of sealing the plate. Alternatively, sealed plates can be stored at 4 °C for up to 4 h before thermal cycling.
9. Only Bio-Rad's AutoDG Pipette tips should be used, as other tips can damage the instrument.
10. The cooling blocks should be cooled in a freezer for at least 2 h before use to prevent droplet/oil emulsion evaporation. Dark purple color indicates that the block is at the correct working temperature. If the block is pink in color, it should not be used.
11. It is recommended to load the oil bottle ahead of time to save time during experiments. To do so, open the instrument, remove the cap from the bottle of AutoDG oil and insert it into the tower of the oil delivery system in the instrument. Touch the oil type button and touch the "Probes" droplet. If loaded correctly, the instrument will perform the flush and prime procedure. Afterwards, touch the "OK" button to return to the home screen. AutoDG Probes oil should be used for hybridization probe-based dPCR experiments.
12. The droplet reader can be switched on in advance to warm up the instrument.
13. Generally, thresholds should be set just above the negative cluster found in NTCs. Sometimes the negative cluster in a given reaction can have a higher fluorescence than the negative cluster in NTCs. The thresholds should be set separately for such reactions.
14. Simplex assays described in the previous subsection can also be adapted for the Stilla platform.
15. The pressure delivered to the instrument should be within 1250 ± 50 mbar and should never exceed 1300 mbar.
16. A new reaction program can be created by clicking "Programs" → "New Program" in the main panel of the instrument software. Modify the program conditions in the created program and then click on the "Save" icon. The programs are saved in the local/scripts folder. Programs can also be imported to or exported from a USB stick.
17. For the 6-color Naica system at least three channels need to be switched on, we recommend the third channel to be the furthest from the detection channels.

Conflict of Interest

The authors have no conflicts of interest to declare that are relevant to the content of this chapter.

References

1. Jones RAC, Koening R, Lesemann DE (1980) Pepino mosaic virus, a new potexvirus from pepino (*Solanum muricatum*). Ann Appl Biol 94:61–68. https://doi.org/10.1111/J.1744-7348.1980.TB03896.X
2. Hančinský R, Mihálik D, Mrkvová M et al (2020) Plant viruses infecting Solanaceae family members in the cultivated and wild environments: a review. Plants 9:667. https://doi.org/10.3390/PLANTS9050667
3. Hanssen IM, Mumford R, Blystad DR et al (2010) Seed transmission of Pepino mosaic virus in tomato. Eur J Plant Pathol 126:145–152. https://doi.org/10.1007/S10658-009-9528-X/TABLES/3
4. Gómez P, Sempere RN, Aranda MA (2012) Pepino mosaic virus and tomato Torrado virus: two emerging viruses affecting tomato crops in the Mediterranean basin. Adv Virus Res 84:505–532. https://doi.org/10.1016/B978-0-12-394314-9.00014-2
5. Moreno-Pérez MG, Pagán I, Aragón-Caballero L et al (2014) Ecological and genetic determinants of Pepino mosaic virus emergence. J Virol 88:3359–3368. https://doi.org/10.1128/JVI.02980-13/ASSET/34409D1C-01AD-4C45-8A2E-1F061DC9565C/ASSETS/GRAPHIC/ZJV9990987770003.JPEG
6. Ling KS, Wechter WP, Jordan R (2007) Development of a one-step immunocapture real-time TaqMan RT-PCR assay for the broad spectrum detection of Pepino mosaic virus. J Virol Methods 144:65–72. https://doi.org/10.1016/J.JVIROMET.2007.03.022
7. Hasiów-Jaroszewska B, Paeleman A, Ortega-Parra N et al (2013) Ratio of mutated versus wild-type coat protein sequences in Pepino mosaic virus determines the nature and severity of yellowing symptoms on tomato plants. Mol Plant Pathol 14:923–933. https://doi.org/10.1111/MPP.12059
8. Management of pepino mosaic virus in greenhouse tomatoes | ontario.ca. https://www.ontario.ca/page/management-pepino-mosaic-virus-greenhouse-tomatoes. Accessed 6 Mar 2024
9. Agüero J, Gómez-Aix C, Sempere RN et al (2018) Stable and broad spectrum cross-protection against Pepino mosaic virus attained by mixed infection. Front Plant Sci 871:425002. https://doi.org/10.3389/FPLS.2018.01810/BIBTEX
10. AgroPages-EU approved emergency use of PMV®-01 vaccine to protect tomato-Agricultural news. https://news.agropages.com/News/NewsDetail%2D%2D-16754-e.htm. Accessed 6 Mar 2024
11. Leibman D, Ortega-Parra N, Wolf D et al (2021) A transgenic RNAi approach for developing tomato plants immune to Pepino mosaic virus. Plant Pathol 70:1003–1012. https://doi.org/10.1111/PPA.13346
12. Córdoba-Sellés MDC, García-Rández A, Alfaro-Fernández A, Jordá-Gutiérrez C (2007) Seed transmission of Pepino mosaic virus and efficacy of tomato seed disinfection treatments. Plant Dis 91:1250–1254. https://doi.org/10.1094/PDIS-91-10-1250
13. Gutiérrez-Aguirre I, Mehle N, Delić D et al (2009) Real-time quantitative PCR based sensitive detection and genotype discrimination of Pepino mosaic virus. J Virol Methods 162:46–55. https://doi.org/10.1016/J.JVIROMET.2009.07.008
14. Alfaro-Fernández A, Sánchez-Navarro JÁ, del Carmen Cebrián M et al (2009) Simultaneous detection and identification of Pepino mosaic virus (PepMV) isolates by multiplex one-step RT-PCR. Eur J Plant Pathol 125:143–158. https://doi.org/10.1007/S10658-009-9466-7/FIGURES/5
15. Sidstedt M, Rådström P, Hedman J (2020) PCR inhibition in qPCR, dPCR and MPS-mechanisms and solutions. Anal Bioanal Chem 412:2009–2023. https://doi.org/10.1007/S00216-020-02490-2
16. Pinheiro LB, Coleman VA, Hindson CM et al (2011) Evaluation of a droplet digital polymerase chain reaction format for DNA copy number quantification. Anal Chem 84:1003–1011. https://doi.org/10.1021/AC202578X
17. Bogožalec Košir A, Demšar T, Štebih D et al (2019) Digital PCR as an effective tool for GMO quantification in complex matrices. Food Chem 294:73–78. https://doi.org/10.1016/J.FOODCHEM.2019.05.029
18. Rački N, Dreo T, Gutierrez-Aguirre I et al (2014) Reverse transcriptase droplet digital PCR shows high resilience to PCR inhibitors from plant, soil and water samples. Plant Methods 10:1–10. https://doi.org/10.1186/S13007-014-0042-6
19. Bogožalec Košir A, Muller S, Žel J et al (2023) Fast and accurate multiplex identification and quantification of seven genetically modified

soybean lines using six-color digital PCR. Foods 12:4156. https://doi.org/10.3390/FOODS12224156/S1

20. Bogožalec Košir A, Spilsberg B, Holst-Jensen A et al (2017) Development and interlaboratory assessment of droplet digital PCR assays for multiplex quantification of 15 genetically modified soybean lines. Sci Rep 7:1–11. https://doi.org/10.1038/s41598-017-09377-w

21. Dobnik D, Spilsberg B, Bogožalec Košir A et al (2015) Multiplex quantification of 12 European Union authorized genetically modified maize lines with droplet digital polymerase chain reaction. Anal Chem 87:8218–8226. https://doi.org/10.1021/ACS.ANALCHEM.5B01208/SUPPL_FILE/AC5B01208_SI_001.PDF

22. Dobnik D, Štebih D, Blejec A et al (2016) Multiplex quantification of four DNA targets in one reaction with Bio-Rad droplet digital PCR system for GMO detection. Sci Rep 6:1–9. https://doi.org/10.1038/srep35451

Chapter 4

Isothermal Amplification Techniques: An Emerging Tool for On-Site Detection of Phytopathogens in Field Conditions

C. Manjunatha, K. Aditya, M. K. Prasannakumar, D. Pramesh, P. Buella, Swathi S. Patil, Gopal Venkateshbabu, J. Harish, R. Karan, and H. V. Ramegowda

Abstract

This chapter reviews various isothermal amplification techniques, which are alternative to PCR for detecting plant pathogens. These methods, including NASBA, SDA, LAMP, HDA, and RPA, amplify nucleic acids at a constant temperature, making them potentially more suitable for point-of-care (POC) applications. The review compares these techniques regarding their mechanisms, advantages, and limitations, highlighting their potential for rapid, cost-effective, and on-site plant disease diagnosis, including the integration of these techniques with CRISPR-Cas systems and lab-on-a-chip technologies.

Key words Isothermal amplification, Plant pathogens, Point-of-care (POC) diagnostics, Nucleic acid amplification, PCR alternative, NASBA (Nucleic acid sequence-based amplification), SDA (Strand displacement amplification), LAMP (Loop-mediated isothermal amplification), HDA (Helicase-dependent amplification), RPA (Recombinase polymerase amplification), CRISPR-Cas, Plant disease detection, Molecular diagnostics

1 Introduction

Plant diseases significantly affect food security and safety. With the global population expected to reach 9.3 billion by 2050, food production must increase by 50% to meet demand [1]. Plant pathogens and insect pests are known to cause up to 40% yield losses in key crops such as maize, rice, and wheat in their growing regions, leading to annual global economic losses estimated at around $220 billion [2]. To mitigate these losses, a crucial factor is the early, accurate, and timely detection of plant pathogens, as it forms the basis for effective plant protection measures in the field [3]. Various sensing methods for plant pathogen detection have been developed and categorized into direct and indirect approaches. Direct methods, like PCR and immunoassays, offer high sensitivity but are

Anthony Torres (ed.), *Genotyping: Methods and Protocols*, Methods in Molecular Biology, vol. 2943,
https://doi.org/10.1007/978-1-0716-4642-7_4,

typically confined to laboratories. Indirect methods, such as thermography and hyperspectral imaging, are non-invasive and useful for early detection. However, these techniques are often time-consuming and expensive and require skilled personnel. While direct methods need complex preparation, indirect methods are limited in large-scale applications due to costly equipment and extensive data analysis [4]. Among all these techniques, nucleic acid amplification methods, especially polymerase chain reaction (PCR), have transformed molecular diagnostics by allowing highly efficient exponential DNA amplification, even from just a single molecule. Despite its advantages, PCR reliance on a thermocycler limits its portability and energy efficiency and has led to the development of alternative nucleic acid amplification methods that operate at a constant temperature and require simpler, energy-efficient equipment [5].

The World Health Organization (WHO) has outlined the essential criteria for point-of-care (POC) testing in their ASSURED guidelines, which emphasize that tests should be affordable, sensitive, specific, user-friendly, robust, rapid, and easy to use and distribute to those in need [6]. Over the past 30 years, numerous isothermal amplification techniques have been developed to meet the ASSURED guidelines basic requirements. These methods are quick, dependable, and can be conducted using a simple water or dry bath with minimal cost. Isothermal amplification techniques utilize strand displacement activity to perform nucleic acid amplification without the need for thermocyclers, thereby removing the temperature cycling step [7]. Hence, several isothermal DNA amplification methods have been developed to meet the amplification needs in low-resource environments or at the POC. There are many isothermal amplification techniques, of which nucleic acid sequence-based amplification (NASBA), strand displacement amplification (SDA), rolling circle amplification (RCA), loop-mediated isothermal amplification (LAMP), helicase-dependent amplification (HDA), and recombinase polymerase amplification (RPA) [8] are important and well known. This chapter reviews different isothermal techniques currently in use and holds potential for in-field direct diagnosis, aiming to simplify the process of plant disease detection for plant protection officials and farmers.

2 Brief History of Isothermal Techniques

The earliest known post-PCR isothermal amplification methods were self-sustained sequence replication (3SR) by Guatelli in 1990 [9]. Between 1991 and 2006, many popular methods like nucleic acid sequence-based amplification (NASBA), strand displacement amplification (SDA), loop-mediated isothermal amplification (LAMP), helicase-dependent amplification (HDA), and

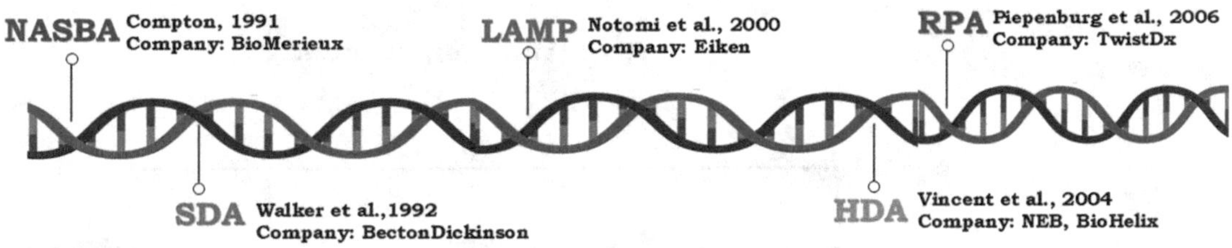

Fig. 1 Timeline illustrating the origin, developers, and commercializing companies of prominent isothermal techniques

recombinase primed amplification (RPA) as depicted in Fig. 1, emerged as key techniques. Currently, advancements in nucleic acid amplification techniques include Hinge-Initiated Primer-Dependent Amplification (HIP) [10] and CRISPR-Cas9-Triggered Nicking Endonuclease-Mediated Strand Displacement Amplification (CRISDA) [11].

3 How Do Isothermal Amplification Techniques Differ from PCR?

Unlike PCR, isothermal amplification is a simple method that amplifies DNA at a constant temperature via the action of specific DNA polymerases, RNA polymerases, and other nucleic acid-modifying enzymes, making the process faster as it eliminates the need for intermittent denaturation steps [7] generating a high copy number in a short time. Isothermal amplification techniques have been extended to detect various targets such as proteins, cells, small molecules, and ions [12]. Integrating isothermal amplification with nanotechnology has also gained attention as these techniques can produce large quantities of nucleic acids, which are useful in biosensing and nanomedicine [7]. Recent advancements in microfabrication have facilitated the integration of isothermal amplification techniques with microfluidic chips, capillary platforms, and paper-based devices. Microfluidic approaches have reduced both amplification reaction times and reaction volumes, enabling digital quantification as an alternative to real-time (kinetic) measurement methods [13].

4 Comparative Analysis of Various Isothermal Amplification Techniques

Each isothermal amplification technique has been individually discussed in this section, and their fundamental characteristics are outlined in Fig. 2.

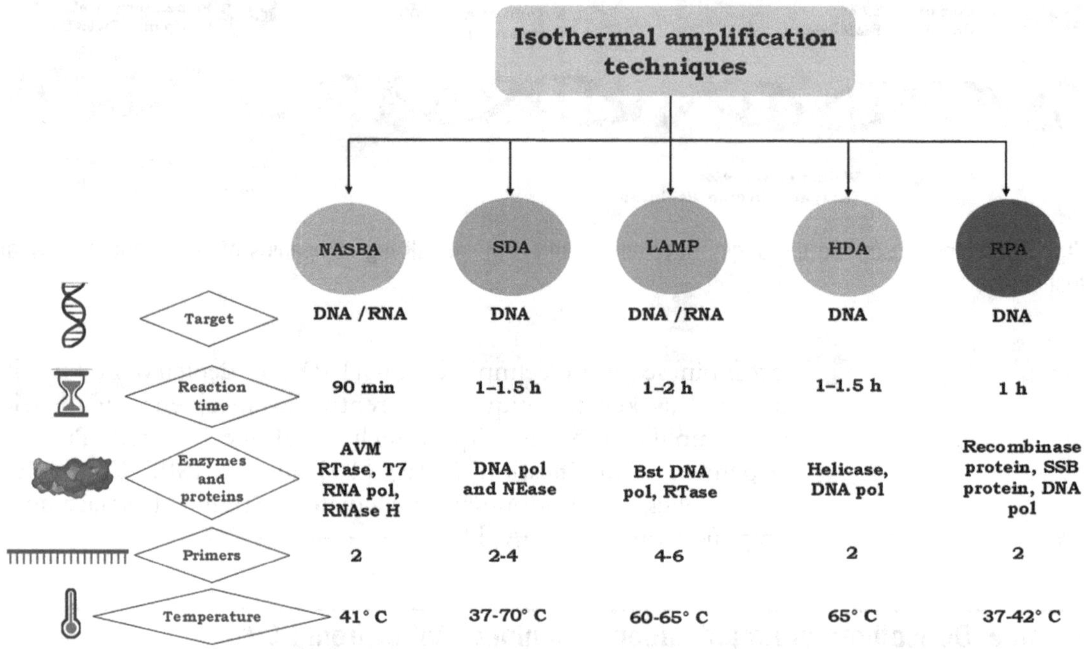

Fig. 2 Comparative analysis of isothermal amplification techniques: targets, reaction time, enzymes and proteins, primers/probes, and temperature

4.1 Nucleic Acid Sequence-Based Amplification (NASBA)

This highly sensitive transcription-based amplification system enables the specific replication of nucleic acids in vitro, operating at a constant temperature of 41 °C. The amplification process involves three key enzymes: avian myeloblastosis virus (AMV) reverse transcriptase (RT), RNase H, and T7 DNA-dependent RNA polymerase (DdRp). Due to the incorporation of RT, this method is particularly well suited for RNA targets like mRNA, rRNA, or genomic RNA [14]. After a brief incubation at 65 °C to denature the target, a forward primer binds to the RNA. This primer not only matches the target sequence but also includes a 5′ sequence for the T7 DdRp promoter. At 41 °C, AMV RT extends the primer, and RNase H degrades the RNA within the RNA-cDNA hybrid. This allows a reverse primer to hybridize and extend to the cDNA, resulting in a double-stranded T7 promoter sequence. The T7 DdRp then synthesizes new RNA molecules complementary to the target RNA. Once the initiation phase is complete, NASBA enters the amplification cycle. The reverse primer binds to newly formed RNA, and RNase H degrades the RNA in the hybrid during its extension. The forward primer then binds to the cDNA, allowing AMV RT to extend and create another double-stranded T7 promoter sequence, leading to the continuous production of new RNA [15]. NASBA-based RNA detection can be performed using various techniques such as the dideoxy method (with reverse transcriptase and labeled

Table 1
List of phytopathogens, targeted genes, and detection techniques for NASBA-based detection

Sl. No.	Target organism	Gene amplified	Detection technique	LOD	References
1	Sugarcane yellow leaf virus (ScYLV)	Coat protein	Fluorescence	100 fg	[17]
2	Strawberry vein banding virus (SVBV)	Coat protein	Coat protein	~1 ng of fresh tissue	[18]
3	*Xanthomonas citri* subsp. *citri* (Xcc),	gumD	Colorimetric, fluorescence	1 fg	[19]
4	Apple scar skin viroid (ASSVd)	–	Colorimetric, fluorescence	10^{12} RNA per μL	[20]

oligonucleotide primer), enzyme-linked gel assays (ELGA), electrochemiluminescent detection, molecular beacons, Southern blot hybridization, and fluorescent correlation spectroscopy [16]. Although NASBA is an established and well-known method, it has not been extensively utilized for the detection of plant pathogens. All detections performed using NASBA are summarized in Table 1.

4.2 Strand Displacement Amplification (SDA)

SDA is an isothermal technique that amplifies DNA without denaturation, relying on strand displacement activity by DNA polymerase and restriction enzymes, and can achieve up to 10^7 fold amplification in 2 h at 37 °C [21, 22]. The process involves three key steps: repeated nicking of target sequences by restriction enzymes, target generation, and exponential amplification. Four primers are used: S1 and S2, which bind to the target sequence with HincII recognition sites, and B1 and B2, which bind solely to target regions, facilitating efficient DNA amplification [21, 22]. Recently, portable devices for Strand Displacement Amplification (SDA) have been developed for healthcare applications [23, 24]. It is highly likely that in the coming years, this technology will also be adapted for point-of-care (POC) testing in the detection of plant pathogens [25].

4.3 Loop-Mediated Isothermal Amplification (LAMP)

LAMP is one of the most widely utilized isothermal amplification technique. Like other isothermal methods, it does not require extensive template or sample preparation, allowing the organism to be directly used as the analyte [26]. The amplification process in LAMP utilizes Bst DNA polymerase, which is derived from *Bacillus stearothermophilus*. This method generates around 10^9 copies within 30–60 min at a constant temperature of 60–65 °C. It involves the use of four specific primers: forward inner primer (FIP), backward inner primer (BIP), forward outer primer (F3), and backward outer primer (B3) [27]. To boost the amplification process, two extra primers, known as forward and reverse loop primers, can be added to the reaction [28]. In the initial phase of LAMP, strand displacement activity forms a dumbbell-shaped structure, while during the secondary phase, cycling amplification occurs, relying solely on two inner primers. This process results in stem-loop DNA with multiple repeats of the target sequence and a cauliflower-like structure containing several loops. The detection efficiency of LAMP can be enhanced by incorporating additional loop primers (LF/LB) [29], so the remarkable specificity of this process has facilitated the use of insoluble pyrophosphate, a reaction by-product in a turbidimetric detection method. This approach can be applied to both qualitative visual detection and real-time quantitative turbidimetry [30]. The LAMP method can also be adapted for RNA detection by incorporating a reverse transcriptase enzyme, which converts RNA into complementary DNA (cDNA). This variant, known as RT-LAMP, relies on reverse transcriptase and Bst polymerase enzymes for amplification [7]. Currently, various methods are employed to detect LAMP amplicons, including gel electrophoresis, naked-eye observation, turbidimetric analysis, colorimetric assays, bioluminescence, fluorescence detection, electrochemical sensors or chips, ELISA, and lateral flow dipsticks (LFD) [31] (Table 2).

4.4 Helicase-Dependent Amplification (HDA)

HDA is an isothermal nucleic acid amplification technique that replicates the natural process of DNA replication by utilizing a helicase enzyme to unwind DNA duplexes at a constant temperature [50]. Besides helicase, the original HDA system requires two additional proteins, MutL and a single-stranded DNA-binding protein (SSB), that help to stabilize the single-stranded DNA (ssDNA), preventing the complementary strands from re-annealing. Once the DNA is unwound and stabilized, two oligonucleotide primers bind to the 5′ and 3′ ends of the target sequence. A DNA polymerase then extends these primers by incorporating deoxynucleotide triphosphates (dNTPs), producing double-stranded amplification products [51]. Previously, the mesophilic form of HDA (mHDA) employed the *Escherichia coli* UvrD helicase paired with the DNA polymerase I Klenow fragment to amplify DNA at 37 °C. Later, the development of a thermostable helicase, Tte-UvrD from *Thermoanaerobacter tengcongenesis*, in combination with *Bacillus stearothermophilus* DNA polymerase (Bst DNA polymerase), created an enzyme called "helimerase" [52], which enabled DNA amplification at higher temperatures (60–65 °C), referred to as thermophilic HDA (tHDA). Unlike mHDA, the tHDA system does not require the MutL protein, simplifying the reaction setup while enhancing the efficiency and selectivity of amplification [53]. HDA is widely utilized in human pathogen diagnostics; however, its application in diagnosing phytopathogens has been relatively limited, as summarized in Table 3.

Table 2
List of phytopathogens, targeted genes, and detection techniques for LAMP-based detection

Sl. No.	Target/organism	Marker/gene	Temperature (°C)	Time (min)	Detection technique	LOD	References
1	*Puccinia triticina*	PtRA 68	65	60	Colorimetric, fluorescence	100 fg/μL	[32]
2	*Magnaporthe oryzae*	RNA polymerase II largest subunit	60	20	Colorimetric	100 fg/μL	[33]
3	*Sarocladium oryzae*	RNA polymerase II largest subunit	60	20	Colorimetric	100 fg/μL	[33]
4	*Xanthomonas axonopodis* pv. *punicae*	–	65	30	Colorimetric, fluorescence	0.153 fg/μL	[34]
5	*Xanthomonas oryzae* pv. *oryzae*	R genes viz., *xa1*, *xa3*, *xa4*, *xa7*, *xa10*, *xa11* and *xa21*	60	30	Colorimetric, fluorescence	100 fg/μL	[35]
6	*Puccinia striiformis f.* sp. *tritici*	Ketopantoate reductase gene	65	60	Colorimetric, fluorescence	1 pg/μL	[36]
7	*Fusarium acuminatum*	Elongation factor-1α (*TEF-1α*)	63	60	Colorimetric, fluorescence	100 fg/μL	[37]
8	*Fusarium solani*	Elongation factor-1α (*TEF-1α*)	62	60	Colorimetric, fluorescence	1 pg/μL	[37]
9	Phytoplasmas	23S rRNA	64	30	Colorimetric, fluorescence	1 pg/μL	[38]
10	*Spongospora subterranea*	rDNA region	60	50	Colorimetric	2 pg/μL	[39]
11	*Fusarium oxysporum* f. sp. *ciceris*	EF-1 alpha	63 °C	60	Colorimetric, fluorescence	100 pg/μL	[40]
12	*Uromyces betae*	Cytochrome *b*	61	<30	Fluorescence	10 pg/μL	[41]
13	*Phytophthora hibernalis*	Enolase	63	60	Colorimetric, fluorescence	0.02 fg	[42]

(continued)

Table 2 (continued)

Sl. No.	Target/organism	Marker/gene	Temperature (°C)	Time (min)	Detection technique	LOD	References
14	*Phytophthora syringae*	Internal transcribed spacers	63	60	Colorimetric, fluorescence	0.02 fg	[42]
15	*Phytophthora cambivora*	ras-like protein *Ypt1*	63	60	Colorimetric, fluorescence	0.02 fg	[42]
16	*Pythium aphanidermatum*	Internal transcribed spacers	68	20	Fluorescence	10 fg	[43]
17	*Phytophthora infestans*	ITS4 and ITS5	65	30	Colorimetric, lateral flow dipstick	1 pg/μL	[44]
18	*Colletotrichum* spp.	β-tubulin 2	64	60	Fluorescence	100 pg/μL	[45]
19	*Fusarium fujikuroi*	Elongation factor 1-α	65	45	–	100 fg/μL	[46]
20	*Xanthomonas gardneri*	*hrpB*	65	30	Colorimetric, fluorescence	1 pg/μL	[47]
21	*Fusarium fujikuroi*	*NRPS31*	65	30	Colorimetric, fluorescence	10 fg/μL	[48]
22	*Tomato leaf curl New Delhi virus*	*CP gene*	65	45	Colorimetric, fluorescence	100 fg/μL	[49]

Table 3
List of phytopathogens, targeted genes, and detection techniques for HDA-based detection

Sl. No.	Target/organism	Marker/gene	Temperature (°C)	Time (min)	Detection technique	LOD	References
1	*Magnaporthe oryzae*	RNA polymerase II largest subunit	55	10	Fluorescence	50 pg/ μL	[33]
2	*Sarocladium oryzae*	RNA polymerase II largest subunit	55	10	Fluorescence	50 pg/ μL	[33]
3	*Phytophthora kernoviae*	Yeast GTP-binding protein (*Ypt1*)	65	60	Fluorescence	10 pg/ μL	[54]
4	High plains virus (HPV)	HPV nucleoprotein	63	200	Fluorescence	1 fg	[55]
5	Tomato spotted wilt virus (TSWV)	N gene	65	60	Fluorescence	4 pg	[56]

4.5 Recombinase Polymerase Amplification (RPA)

RPA is a very easy-to-use approach and is becoming increasingly favored for point-of-care (POC) analysis of phytopathogens due to its numerous advantages over PCR and even LAMP [25]. The core reaction mechanism of RPA is based on an engineered adaptation of the natural cellular process known as homologous recombination, which plays a crucial role in DNA metabolism. The standard RPA reaction involves three primary proteins, recombinase, a recombinase loading factor, and single-stranded binding proteins, that work together with additional components like DNA polymerase, a molecular crowding agent, energy sources such as adenosine triphosphate (ATP), and salts to facilitate the RPA process [57]. The RPA process begins when the uvsX recombinase protein, derived from T4-like bacteriophages, binds to primers in the presence of ATP and a crowding agent such as high molecular weight polyethylene glycol. This forms a recombinase-primer complex, which scans double-stranded DNA to locate a homologous sequence, enabling strand invasion by the primer at the complementary site. Single-stranded binding proteins stabilize the displaced DNA strand to prevent primer displacement through branch migration. Once the recombinase dissociates, a strand-displacing DNA polymerase, such as the large fragment of *Bacillus subtilis* Pol1 (Bsu), binds to the primer's 3′ end and extends it with the help of dNTPs. Through continuous cycling, this process results in exponential DNA amplification [8]. It has been coupled with FRET-based fluorescence probes and lateral flow strip-sensing systems to minimize background noise and enable instrument-free detection [7]. Since 2012, RPA-based tests for various pathogens have significantly developed. Specialized commercial kits utilizing RPA have been developed to detect various plant pathogens, including the plum pox virus, grapevine red blotch-associated virus, *Candidatus Liberibacter asiaticus*, *Fusarium oxysporum* f. sp. *vasinfectum* (Race 4), little cherry virus 2, banana bunchy top virus, and tomato chlorotic dwarf viroid. These kits feature basic equipment for homogenization (such as buffer, plastic bags, and mesh), a lyophilized premix designed for the RPA reaction that includes optimized primers and a tetrahydrofuran probe, as well as a lateral flow test cassette. The first RT-RPA test for little cherry virus was developed in 2014 [58]. Currently, RPA is the only isothermal technique having commercially available kits specifically designed for plant pathogen detection [58]. RPA applied for detecting numerous phytopathogens is enlisted in Table 4.

5 Advantages and Limitations of Isothermal Amplification Techniques

Although isothermal amplification techniques offer numerous advantages, they also have specific drawbacks and limitations that restrict their applicability under certain conditions (Table 5).

6 Conclusion and Prospect for the Future

Early and on-site detection of plant pathogens can potentially lead to effective crop disease management [34]. On-site plant pathogens detection requires simple, quick steps for processing plant tissues, extracting nucleic acids, amplifying, and detecting. Currently, portable and easy-to-use equipment, kits, and protocols based on isothermal amplification techniques are particularly suited for on-site diagnostics due to their specific features [58].

While PCR is the gold standard in nucleic acid testing, isothermal nucleic acid amplification has emerged as a transformative alternative [57]. Isothermal methods maintain a constant, low-temperature profile and can utilize various types of starting materials. They are less sensitive to inhibitors and provide a broad array of detection options, making them ideal for simpler integration into lab-on-a-chip systems. This ease of integration boosts their implementation and growing importance. Additionally, incorporating these methods into chip formats enhances sensitivity, reproducibility, speed, and cost-effectiveness while improving result accuracy and reducing amplification bias. These combined advantages are significant, challenging PCR for its status as the "gold standard" [72].

Table 4
List of phytopathogens, targeted genes, and detection techniques for RPA-based detection

Sl. No.	Target/organism	Marker/gene	Temperature (°C)	Time (min)	Detection technique	LOD	References
1	*Botryosphaeria dothidea*	β-Tubulin (*TUB*)	39	5	Fluorescence	2 pg	[59]
2	*Diaporthe eres*	Histone H3 (*HIS*)	39	5	Fluorescence	2 pg	[59]
3	*Fusarium fujikuroi*	*cyp51c*	39	8	Lateral flow device (LFD)	100 fg/μL	[60]
4	Plum pox virus (PPV)	Coat protein (CP)	42	10	Fluorescence	10 pg/μL	[61]
5	*Phytophthora sojae*	*atp9-nad9* region of the mitochondrial genome	39	25	Fluorescence	10 pg	[62]
6	Sugarcane bacilliform virus (SCBV)	–	65	10	–	10 ag/μL	[63]
7	Citrus tristeza virus (CTV)	–	38	30	Colorimetric, fluorescence	0.01 fg/μL	[64]
8	*Phytophthora infestans*	ITS	39	30	Fluorescence	50 fg/μL	[65]
9	*Dickeya solani*	SOL-C genomic region	39	30	Fluorescence	1.4×10^4 CFU per gram of potato tuber	[66]
10	Potato virus X	Coat protein gene (gp5)	37	20	Fluorescence	0.14 pg virus per gram of plant leaf	[67]
11	Potato spindle tuber viroid	–	39	20	Fluorescence	10^6 copies of in vitro transcribed PSTV RNA	[68]
12	Tomato spotted wilt virus	Coat protein	38	10	Lateral flow strips (LFS)	10 fg/μL of transcribed TSWV RNA	[69]
13	Citrus tristeza virus	Coat protein	40	20	Fluorescence	$\leq$0.23 ng/μL of CTV RNA	[70]
14	Milk vetch dwarf virus	Coat protein	37	30	Fluorescence	10^1 copies/μL	[71]

Table 5
Primary isothermal amplification systems, emphasizing their main advantages and disadvantages

Sl. No.	Isothermal technique	Advantages	Limitations	References
1	NASBA	NASBA can detect multiple pathogens simultaneously It can amplify target RNA directly It produces a high copy number at a constant temperature and is commercially available in a compact device	It requires costly enzymes that need to be added individually It necessitates a high-temperature denaturation step It is less effective at amplifying longer RNA targets	[7, 58]
2	SDA	Suitable for miRNA profiling Power saving	A high-temperature initial denaturation step is necessary Limited to small targets Prone to unspecific amplification	[72]
3	LAMP	The equipment and reaction components are inexpensive	An accurate primer design and many primers are needed for selection It is not well-suited for amplifying small fragments of target nucleic acids The primer design process is complex, typically requiring four to six primers, with two of the primers in the set being longer, leading to increased synthesis costs Complex multiplexing	[7, 73]
4	HDA	It features a simple reaction scheme where a target sequence can be amplified using two flanking primers, much like in PCR HDA is a true isothermal DNA amplification method	HDA is quite slow	[58]
5	RPA	A very low temperature is needed, as reactions can be initiated even by body heat A very brief amplification time of 10–30 min Real-time detection is possible in a single tube	RPA has a limitation of 1500 base pairs on the length of the amplification product The ideal length of the RPA DNA product for a lateral flow assay (LFA) is 150 base pairs RPA can produce false positive results when there are up to nine mismatches present in both primers The high viscosity of the reaction mixture reduces diffusion, which can impact the local amplification process and the volume of amplification	[74]

Several factors should be considered before selecting an isothermal amplification method for a specific application, especially the sample type. Complex samples with inhibitory components may require purification beforehand. If the template concentration is high, diluting the sample alongside a robust polymerase can help reduce inhibition. In cases where nucleic acids are enclosed within viral particles or rigid cells, a lysis step is necessary to access the template. Commercial kits are widely available for sample purification.

Additionally, for RNA templates, it's crucial to eliminate or inhibit RNases and promptly proceed with the amplification reaction [5]. The nonspecific amplification is obvious in isothermal amplification due to the compromised stringency of primer binding. Several additives such as polyethylene glycol, dextran, proline, betaine, and helicases have been used to enhance the specificity of isothermal amplification methods. The coupling of isothermal amplification techniques with methods such as colorimetric, nanoparticles, lateral flow biosensors, etc., leads to the development of novel sensing techniques. However, some techniques need to be optimized to produce better results and increase their use at POC sites for diagnosis. Notably, each amplification technique has its specialty, which increases when coupled with other methods [7].

Interestingly, emerging sequencing technologies are now seen as the primary competitor to isothermal amplification assays [75]. Portable devices like the Oxford Nanopore MinION are already enabling flexible bedside diagnostics by allowing rapid sequencing and identification of antibiotic resistance genes [76] or performing multiplex pharmacogenetics tests within a few hours [77]. Nevertheless, in many contexts, the affordability and speed offered by isothermal amplification assays will remain highly advantageous [5].

However, the broader success of isothermal methods depends on further validation of their specificity and cost-effectiveness. Many of these techniques are prone to producing nonspecific products and high background noise due to the low amplification temperatures. These challenges need to be resolved before clinical implementation. Additionally, numerous methods are still under patent or licensed to single companies, which restricts access to key enzymes and reagents, resulting in higher market prices and limiting widespread use. Once these patents expire, isothermal amplification techniques are expected to meet nucleic acid diagnostics and point-of-care application standards, leading to more profitable global commercialization [72]. Isothermal techniques coupled with CRISPR-Cas are the future of diagnostics in the days to come as they are more accurate than traditional isothermal techniques.

Conflict of Interest

The authors have no conflicts of interest to declare that are relevant to the content of this chapter.

References

1. Alexandratos N, Bruinsma J (2012) World agriculture towards 2030/2050: the 2012 revision. FAO
2. Wang N, Sundin GW, Fuente L, Cubero J, Tatineni S, Brewer MT, Zeng Q, Bock CH, Cunniffe NJ, Wang C, Candresse T, Chappell T, Coleman JJ, Munkvold G (2024) Key challenges in plant pathology in the next decade. Phytopathology 114(5):837–842. https://doi.org/10.1094/PHYTO-04-24-0137-KC
3. Gautam AK, Kumar S (2020) Techniques for the detection, identification, and diagnosis of agricultural pathogens and diseases. In: Natural remedies for pest, disease and weed control. Elsevier, pp 135–142. https://doi.org/10.1016/b978-0-12-819304-4.00012-9
4. Patel R, Mitra B, Vinchurkar M, Adami A, Patkar R, Giacomozzi F, Lorenzelli L, Baghini MS (2022) A review of recent advances in plant-pathogen detection systems. Heliyon 8(12):e11855. https://doi.org/10.1016/j.heliyon.2022.e11855
5. Glokler J, Lim TS, Ida J, Frohme M (2021) Isothermal amplifications—a comprehensive review on current methods. Crit Rev Biochem Mol Biol 56(6):543–586. https://doi.org/10.1080/10409238.2021.1937927
6. Martzy R, Kolm C, Krska R, Mach RL, Farnleitner AH, Reischer GH (2019) Challenges and perspectives in the application of isothermal DNA amplification methods for food and water analysis. Anal Bioanal Chem 411(9):1695–1702. https://doi.org/10.1007/s00216-018-1553-1
7. Srivastava P, Prasad D (2023) Isothermal nucleic acid amplification and its uses in modern diagnostic technologies. 3 Biotech 13(6):200. https://doi.org/10.1007/s13205-023-03628-6
8. Lobato IM, O'Sullivan CK (2018) Recombinase polymerase amplification: basics, applications and recent advances. Trends Analyt Chem 98:19–35. https://doi.org/10.1016/j.trac.2017.10.015
9. Guatelli JC, Whitfield KM, Kwoh DY, Barringer KJ, Richman DD, Gingeras TR (1990) Isothermal, in vitro amplification of nucleic acids by a multienzyme reaction modeled after retroviral replication. Proc Natl Acad Sci USA 87(5):1874–1878. https://doi.org/10.1073/pnas.87.5.1874
10. Fischbach J, Frohme M, Glokler J (2017) Hinge-initiated primer-dependent amplification of nucleic acids (HIP)—a new versatile isothermal amplification method. Sci Rep 7(1):7683. https://doi.org/10.1038/s41598-017-08067-x
11. Zhou W, Hu L, Ying L, Zhao Z, Chu PK, Yu XF (2018) A CRISPR-Cas9-triggered strand displacement amplification method for ultrasensitive DNA detection. Nat Commun 9(1):5012. https://doi.org/10.1038/s41467-018-07324-5
12. Niemz A, Ferguson TM, Boyle DS (2011) Point-of-care nucleic acid testing for infectious diseases. Trends Biotechnol 29(5):240–250. https://doi.org/10.1016/j.tibtech.2011.01.007
13. Zhang H, Xu Y, Fohlerova Z, Chang H, Iliescu C, Neuzil P (2019) LAMP-on-a-chip: revising microfluidic platforms for loop-mediated DNA amplification. Trends Analyt Chem 113:44–53. https://doi.org/10.1016/j.trac.2019.01.015
14. Fakruddin M, Mazumdar RM, Chowdhury A, Mannan KB (2012) Nucleic acid sequence based amplification (NASBA)—prospects and applications. Int J Life Sci Pharma Res 2(1):106–121
15. Deiman B, van Aarle P, Sillekens P (2002) Characteristics and applications of nucleic acid sequence-based amplification (NASBA). Mol Biotechnol 20(2):163–179. https://doi.org/10.1385/MB:20:2:163
16. Compton J (1991) Nucleic acid sequence-based amplification. Nature 350(6313):91–92. https://doi.org/10.1038/350091a0
17. Gonçςalves MC, Klerks MM, Verbeek M, Vega JV, Van den Heuvel JF (2002) The use of molecular beacons combined with NASBA for the sensitive detection of sugarcane yellow leaf virus. Eur J Plant Pathol 108:401–407
18. Vašková D, Špak J, Klerks MM, Schoen CD, Thompson JR, Jelkmann W (2004) Real-time NASBA for detection of strawberry vein banding virus. Eur J Plant Pathol 110:213–221

19. Scuderi G, Golmohammadi M, Cubero J, López MM, Cirvilleri G, Llop P (2010) Development of a simplified NASBA protocol for detecting viable cells of the citrus pathogen Xanthomonas citri subsp. citri under different treatments. Plant Pathol 59(4):764–772. https://doi.org/10.1111/j.1365-3059.2010.02305.x
20. Heo S, Kim HR, Lee HJ (2019) Development of a quantitative real-time nucleic acid sequence based amplification (NASBA) assay for early detection of apple scar skin viroid. Plant Pathol J 35(2):164–171. https://doi.org/10.5423/PPJ.OA.10.2018.0206
21. Walker GT, Fraiser MS, Schram JL, Little MC, Nadeau JG, Malinowski DP (1992) Strand displacement amplification—an isothermal, in vitro DNA amplification technique. Nucleic Acids Res 20(7):1691–1696. https://doi.org/10.1093/nar/20.7.1691
22. Walker GT, Little MC, Nadeau JG, Shank DD (1992) Isothermal in vitro amplification of DNA by a restriction enzyme/DNA polymerase system. Proc Natl Acad Sci USA 89(1):392–396. https://doi.org/10.1073/pnas.89.1.392
23. Song J, Liu C, Mauk MG, Peng J, Schoenfeld T, Bau HH (2018) A multifunctional reactor with dry-stored reagents for enzymatic amplification of nucleic acids. Anal Chem 90(2):1209–1216. https://doi.org/10.1021/acs.analchem.7b03834
24. Venzac B, Diakite ML, Herthnek D, Cisse I, Bockelmann U, Descroix S, Malaquin L, Viovy JL (2017) On-chip conductometric detection of short DNA sequences via electro-hydrodynamic aggregation. Analyst 143(1):190–199. https://doi.org/10.1039/c7an00798a
25. Baldi P, La Porta N (2020) Molecular approaches for low-cost point-of-care pathogen detection in agriculture and forestry. Front Plant Sci 11:570862. https://doi.org/10.3389/fpls.2020.570862
26. Zanoli LM, Spoto G (2013) Isothermal amplification methods for the detection of nucleic acids in microfluidic devices. Biosensors (Basel) 3(1):18–43. https://doi.org/10.3390/bios3010018
27. Notomi T, Okayama H, Masubuchi H, Yonekawa T, Watanabe K, Amino N, Hase T (2000) Loop-mediated isothermal amplification of DNA. Nucleic Acids Res 28(12):E63. https://doi.org/10.1093/nar/28.12.e63
28. Nagamine K, Hase T, Notomi T (2002) Accelerated reaction by loop-mediated isothermal amplification using loop primers. Mol Cell Probes 16(3):223–229. https://doi.org/10.1006/mcpr.2002.0415
29. Notomi T, Mori Y, Tomita N, Kanda H (2015) Loop-mediated isothermal amplification (LAMP): principle, features, and future prospects. J Microbiol 53(1):1–5. https://doi.org/10.1007/s12275-015-4656-9
30. Yang Q, Zhang Y, Li C, Zhao Y, Ma X, Zhang W, Zhang S (2018) A rapid and visual single primer isothermal amplification-based method for the detection of Staphylococcus aureus in raw pork products. Food Anal Methods 11(11):3113–3120. https://doi.org/10.1007/s12161-018-1294-y
31. Zhong J, Zhao X (2018) Isothermal amplification technologies for the detection of foodborne pathogens. Food Anal Methods 11(6):1543–1560. https://doi.org/10.1007/s12161-018-1177-2
32. Manjunatha C, Sharma S, Kulshreshtha D, Gupta S, Singh K, Bhardwaj SC, Aggarwal R (2018) Rapid detection of Puccinia triticina causing leaf rust of wheat by PCR and loop mediated isothermal amplification. PLoS One 13(4):e0196409. https://doi.org/10.1371/journal.pone.0196409
33. Prasannakumar MK, Parivallal PB, Pramesh D, Mahesh HB, Raj E (2021) LAMP-based foldable microdevice platform for the rapid detection of Magnaporthe oryzae and Sarocladium oryzae in rice seed. Sci Rep 11(1):178. https://doi.org/10.1038/s41598-020-80644-z
34. Prasannakumar MK, Buela Parivallal P, Manjunatha C, Mahesh HB, Pramesh D, Narayan KS, Gopal VB, Priyanka K, Puneeth ME, Rangaswamy KT (2020) Loop-mediated isothermal amplification assay for pre-symptomatic stage detection of Xanthomonas axonopodis pv. punicae infection in pomegranate. Australas Plant Pathol 49(4):467–473. https://doi.org/10.1007/s13313-020-00720-w
35. Prasannakumar MK, Parivallal BP, Manjunatha C, Pramesh D, Narayan KS, Venkatesh G, Banakar SN, Mahesh HB, Vemanna RS, Rangaswamy KT (2021) Rapid genotyping of bacterial leaf blight resistant genes of rice using loop-mediated isothermal amplification assay. Mol Biol Rep 48(1):467–474. https://doi.org/10.1007/s11033-020-06077-z
36. Aggarwal R, Sharma S, Manjunatha C, Gupta S, Singh VK (2017) Development and validation of loop mediated isothermal amplification based detection assay for Puccinia striiformis f. sp. tritici causing stripe rust of wheat. Australas Plant Pathol 46(6):577–583. https://doi.org/10.1007/s13313-017-0524-x
37. Wang Y, Wang C, Ma Y, Zhang X, Yang H, Li G, Li X, Wang M, Zhao X, Wang J, Zhang X

(2022) Rapid and specific detection of Fusarium acuminatum and Fusarium solani associated with root rot on Astragalus membranaceus using loop-mediated isothermal amplification (LAMP). Eur J Plant Pathol 163(2):305–320. https://doi.org/10.1007/s10658-022-02478-x

38. Akahori M, Miyazaki A, Koinuma H, Tokuda R, Iwabuchi N, Kitazawa Y, Maejima K, Namba S, Yamaji Y (2024) Use of the 23S rRNA gene as a target template in the universal loop-mediated isothermal amplification (LAMP) of genomic DNA from phytoplasmas. Microbiol Spectr 12(5):e0010624. https://doi.org/10.1128/spectrum.00106-24
39. Jiang J, Feindel W, Harding M, Feindel D, Bajema S, Feng J (2023) Development and evaluation of a loop-mediated isothermal amplification (LAMP) method for detection of the potato powdery scab pathogen Spongospora subterranea. Plant Dis 107(1):136–141. https://doi.org/10.1094/PDIS-05-22-1000-SR
40. Ghosh R, Nagavardhini A, Sengupta A, Sharma M (2015) Development of loop-mediated isothermal amplification (LAMP) assay for rapid detection of Fusarium oxysporum f. sp. ciceris—wilt pathogen of chickpea. BMC Res Notes 8:40. https://doi.org/10.1186/s13104-015-0997-z
41. Kaczmarek AM, King KM, West JS, Stevens M, Sparkes D, Dickinson MJ (2019) A loop-mediated isothermal amplification (LAMP) assay for rapid and specific detection of airborne inoculum of Uromyces betae (sugar beet rust). Plant Dis 103(3):417–421. https://doi.org/10.1094/PDIS-02-18-0337-RE
42. Li G-R, Huang G-M, Zhu L-H, Lv D, Cao B, Liao F, Luo J-F (2018) Loop-mediated isothermal amplification (LAMP) detection of Phytophthora hibernalis, P. syringae and P. cambivora. J Plant Pathol 101(1):51–57. https://doi.org/10.1007/s42161-018-0136-5
43. Fukuta S, Takahashi R, Kuroyanagi S, Miyake N, Nagai H, Suzuki H, Hashizume F, Tsuji T, Taguchi H, Watanabe H, Kageyama K (2013) Detection of Pythium aphanidermatum in tomato using loop-mediated isothermal amplification (LAMP) with species-specific primers. Eur J Plant Pathol 136(4):689–701. https://doi.org/10.1007/s10658-013-0198-3
44. Ristaino JB, Saville AC, Paul R, Cooper DC, Wei Q (2020) Detection of Phytophthora infestans by loop-mediated isothermal amplification, real-time LAMP, and droplet digital PCR. Plant Dis 104(3):708–716. https://doi.org/10.1094/PDIS-06-19-1186-RE
45. Liu Y, Ji Y, Han Y, Song L, Zhang L, Ning Z, Yan W, Gao Q, Duan K (2021) Loop-mediated isothermal amplification and PCR combined assay to detect and distinguish latent Colletotrichum spp. infection on strawberry. J Plant Pathol 103(3):887–899. https://doi.org/10.1007/s42161-021-00873-7
46. Sanna M, Spadaro D, Gullino ML, Mezzalama M (2021) Optimization of a loop-mediated isothermal amplification assay for on-site detection of Fusarium fujikuroi in rice seed. Agronomy 11(8):1580. https://doi.org/10.3390/agronomy11081580
47. Stehlikova D, Beran P, Cohen SP, Curn V (2020) Development of real-time and colorimetric loop mediated isothermal amplification assay for detection of Xanthomonas gardneri. Microorganisms 8(9):1301. https://doi.org/10.3390/microorganisms8091301
48. Sunani SK, Bashyal BM, Rawat K, Manjunatha C, Sharma S, Prakash G, Krishnan SG, Singh AK, Aggarwal R (2019) Development of PCR and loop mediated isothermal amplification assay for the detection of bakanae pathogen Fusarium fujikuroi. Eur J Plant Pathol 154(3):715–725. https://doi.org/10.1007/s10658-019-01694-2
49. Naganur P, Premchand U, Shankarappa KS, Mesta RK, Manjunatha C, Patil CV (2019) Development of a loop-mediated isothermal amplification assay for detection of tomato leaf curl New Delhi virus in ridge gourd [Luffa acutangula (L) Roxb.]. Int J Curr Microbiol App Sci 8(09):2282–2295. https://doi.org/10.20546/ijcmas.2019.809.264
50. Vincent M, Xu Y, Kong H (2004) Helicase-dependent isothermal DNA amplification. EMBO Rep 5(8):795–800. https://doi.org/10.1038/sj.embor.7400200
51. Barreda-Garcia S, Miranda-Castro R, de-Los--Santos-Alvarez N, Miranda-Ordieres AJ, Lobo-Castanon MJ (2018) Helicase-dependent isothermal amplification: a novel tool in the development of molecular-based analytical systems for rapid pathogen detection. Anal Bioanal Chem 410(3):679–693. https://doi.org/10.1007/s00216-017-0620-3
52. Motre A, Li Y, Kong H (2008) Enhancing helicase-dependent amplification by fusing the helicase with the DNA polymerase. Gene 420(1):17–22. https://doi.org/10.1016/j.gene.2008.04.017
53. An L, Tang W, Ranalli TA, Kim HJ, Wytiaz J, Kong H (2005) Characterization of a

thermostable UvrD helicase and its participation in helicase-dependent amplification. J Biol Chem 280(32):28952–28958. https://doi.org/10.1074/jbc.M503096200

54. Schwenkbier L, Pollok S, Rudloff A, Sailer S, Cialla-May D, Weber K, Popp J (2015) Non-instrumented DNA isolation, amplification and microarray-based hybridization for a rapid on-site detection of devastating Phytophthora kernoviae. Analyst 140(19): 6610–6618. https://doi.org/10.1039/c5an00855g
55. Arif M, Aguilar-Moreno GS, Wayadande A, Fletcher J, Ochoa-Corona FM (2014) Primer modification improves rapid and sensitive in vitro and field-deployable assays for detection of high plains virus variants. Appl Environ Microbiol 80(1):320–327. https://doi.org/10.1128/AEM.02340-13
56. Wu X, Chen C, Xiao X, Deng MJ (2016) Development of reverse transcription thermostable helicase-dependent DNA amplification for the detection of tomato spotted wilt virus. J AOAC Int 99(6):1596–1599. https://doi.org/10.5740/jaoacint.16-0132
57. Li J, Macdonald J, von Stetten F (2020) Correction: Review: a comprehensive summary of a decade development of the recombinase polymerase amplification. Analyst 145(5): 1950–1960. https://doi.org/10.1039/c9an90127b
58. Ivanov AV, Safenkova IV, Zherdev AV, Dzantiev BB (2021) The potential use of isothermal amplification assays for in-field diagnostics of plant pathogens. Plants (Basel) 10(11):2424. https://doi.org/10.3390/plants10112424
59. Park GG, Kim W, Yang KY (2023) Rapid and sensitive detection of the causal agents of postharvest kiwifruit rot, Botryosphaeria dothidea and Diaporthe eres, using a recombinase polymerase amplification assay. Plant Pathol J 39(5):522–527. https://doi.org/10.5423/PPJ.NT.07.2023.0094
60. Zhang F, Jin C, Hu R, Li Z, Hu S, Zhang Y, Zhang C (2023) Rapid detection of Fusarium fujikuroi in rice seeds and soaking water samples based on recombinase polymerase amplification-lateral flow dipstick. Plant Pathol 73(3):590–601. https://doi.org/10.1111/ppa.13831
61. Jeong H-W, Lee H-J, Cho I-S, Ju H-J, Jeong R-D (2021) Rapid detection of plum pox virus by reverse transcription recombinase polymerase amplification. J Plant Dis Protect 128(3): 881–885. https://doi.org/10.1007/s41348-021-00452-z
62. Rojas JA, Miles TD, Coffey MD, Martin FN, Chilvers MI (2017) Development and application of qPCR and RPA genus- and species-specific detection of Phytophthora sojae and P. sansomeana root rot pathogens of soybean. Plant Dis 101(7):1171–1181. https://doi.org/10.1094/PDIS-09-16-1225-RE
63. Parameswari BSS, Karthikaiselvi L, Wangkhem B, Nithya K, Viswanathan R, Rao GP, Chalam VC, Anitha K (2023) A simplified isothermal recombinase polymerase amplification assay for detection of sugarcane bacilliform virus infection. Sugar Tech 25:995–999
64. Sharma SK, Pathaw N, Wangkhem B, Jackson KS, Devi KS, Roy SS, Singh AR, Singh R, Banerjee A, Kumar S, Ningombam A, Firake DM, Singh TS (2023) Simple template-based reverse transcription-recombinase polymerase amplification assay for routine diagnosis of citrus tristeza virus. Lett Appl Microbiol 76(1): ovac060. https://doi.org/10.1093/lambio/ovac060
65. Si Ammour M, Bilodeau GJ, Tremblay DM, Van der Heyden H, Yaseen T, Varvaro L, Carisse O (2017) Development of real-time isothermal amplification assays for on-site detection of Phytophthora infestans in potato leaves. Plant Dis 101(7):1269–1277. https://doi.org/10.1094/PDIS-12-16-1780-RE
66. Ivanov AV, Safenkova IV, Drenova NV, Zherdev AV, Dzantiev BB (2020) Development of lateral flow assay combined with recombinase polymerase amplification for highly sensitive detection of Dickeya solani. Mol Cell Probes 53:101622. https://doi.org/10.1016/j.mcp.2020.101622
67. Ivanov AV, Safenkova IV, Zherdev AV, Dzantiev BB (2020) Nucleic acid lateral flow assay with recombinase polymerase amplification: solutions for highly sensitive detection of RNA virus. Talanta 210:120616. https://doi.org/10.1016/j.talanta.2019.120616
68. Ivanov AV, Shmyglya IV, Zherdev AV, Dzantiev BB, Safenkova IV (2020) The challenge for rapid detection of high-structured circular RNA: assay of potato spindle tuber viroid based on recombinase polymerase amplification and lateral flow tests. Plants (Basel) 9(10):1369. https://doi.org/10.3390/plants9101369
69. Lee HJ, Cho IS, Ju HJ, Jeong RD (2021) Rapid and visual detection of tomato spotted wilt virus using recombinase polymerase amplification combined with lateral flow strips. Mol Cell Probes 57:101727. https://doi.org/10.1016/j.mcp.2021.101727
70. Ghosh DK, Kokane SB, Gowda S (2020) Development of a reverse transcription recombinase polymerase based isothermal

amplification coupled with lateral flow immunochromatographic assay (CTV-RT-RPA-LFICA) for rapid detection of Citrus tristeza virus. Sci Rep 10(1):20593. https://doi.org/10.1038/s41598-020-77692-w

71. Cao Y, Yan D, Wu X, Chen Z, Lai Y, Lv L, Yan F, Chen J, Zheng H, Song X (2020) Rapid and visual detection of milk vetch dwarf virus using recombinase polymerase amplification combined with lateral flow strips. Virol J 17(1):102. https://doi.org/10.1186/s12985-020-01371-5

72. Oliveira BB, Veigas B, Baptista PV (2021) Isothermal amplification of nucleic acids: the race for the next "gold standard". Front Sensors 2: 752600. https://doi.org/10.3389/fsens.2021.752600

73. Shivakumar N, Nanjundaiah SA, Thiruvengadam V, Channappa M, Thammayya SK, Aravindaram K, Sushil SN (2024) Development of recombinase polymerase amplification-based colorimetric detection assay for rapid identification of invasive cassava mealybug, Phenacoccus manihoti Matile-Ferrero. Saudi J Biol Sci 31(6):104005. https://doi.org/10.1016/j.sjbs.2024.104005

74. Nanditha S, Manjunatha C, Shivakumara KT, Ramya RS, Kandan A, Prasannakumar MK, Pramesh D, Sushil SN (2024) Development of point-of-need colourimetric, isothermal diagnostic assays for specific detection of Bacillus subtilis using shikimate dehydrogenase gene. Folia Microbiol (Praha). https://doi.org/10.1007/s12223-024-01201-z

75. Duncan R, Grigorenko E, Fisher C, Hockman D, Lanning B (2019) Advances in multiplex nucleic acid diagnostics for blood-borne pathogens: promises and pitfalls—an update. Expert Rev Mol Diagn 19(1):15–25. https://doi.org/10.1080/14737159.2019.1559055

76. Brinda K, Callendrello A, Ma KC, MacFadden DR, Charalampous T, Lee RS, Cowley L, Wadsworth CB, Grad YH, Kucherov G, O'Grady J, Baym M, Hanage WP (2020) Rapid inference of antibiotic resistance and susceptibility by genomic neighbour typing. Nat Microbiol 5(3):455–464. https://doi.org/10.1038/s41564-019-0656-6

77. Liau Y, Cree SL, Maggo S, Miller AL, Pearson JF, Gladding PA, Kennedy MA (2019) A multiplex pharmacogenetics assay using the MinION nanopore sequencing device. Pharmacogenet Genomics 29(9):207–215. https://doi.org/10.1097/FPC.0000000000000385

Chapter 5

Loop-Mediated Isothermal Amplification (LAMP) Genotyping for Presence–Absence Determination of Gene of Interest in Test Samples

Anthony Torres and Reginald Gaudino

Abstract

This chapter explores the use of loop-mediated isothermal amplification (LAMP) for rapid and cost-effective genotyping, highlighting its advantages in field-based or resource-limited settings due to its isothermal nature. DNA is extracted from FTA samples and leaf tissue for input into a downstream LAMP assay targeting specific genes. The method's high sensitivity and specificity enables accurate detection of genotypes with simple colorimetric analysis. While challenges exist in multiplexing and non-specific amplification, this protocol highlights LAMP's simplicity and efficiency, demonstrating it is a valuable tool for plant diagnostics and breeding.

Key words Loop-mediated isothermal amplification (LAMP), Genotyping, Presence/absence determination, Plant diagnostics, Agricultural biotechnology, DNA, Amplicon, Isothermal

1 Introduction

In agricultural biotechnology, rapid and accurate identification of specific genetic traits or the presence of pathogens is crucial for effective crop management and breeding programs [1, 2]. Loop-mediated isothermal amplification (LAMP), a molecular diagnostic technique, has emerged as an accurate and efficient molecular tool [3]. LAMP's unique isothermal nature eliminates the need for complex and expensive thermal cycling equipment, traditionally required for polymerase chain reaction (PCR) [4]. This characteristic significantly reduces costs and simplifies the process, making LAMP particularly suitable for field-based or resource-limited applications [5]. Moreover, LAMP is recognized for its sensitivity and specificity, enabling the accurate detection of target DNA of genes observed in a presence–absence distribution such as sex determination in plants [6]. The resulting amplified products can often be visualized through simple colorimetric analysis, enhancing

Anthony Torres (ed.), *Genotyping: Methods and Protocols*, Methods in Molecular Biology, vol. 2943, https://doi.org/10.1007/978-1-0716-4642-7_5,

the method's accessibility and practicality [7]. Furthermore, in plant breeding programs, LAMP enables marker-assisted selection, accelerating the development of desired traits [8].

While LAMP genotyping offers robust advantages, it is not without its challenges [9]. Multiplexing, or the simultaneous detection of multiple targets, is limited by single target analysis, and assay development can be complex and requires careful optimization [10]. Ongoing research and development efforts are focused on overcoming these limitations and expanding the use cases and capabilities of LAMP [11]. Routinely, LAMP genotyping in agricultural biotechnology is a field-deployable solution for presence/absence determination in target genotypes [6]. Its simplicity, sensitivity, and specificity make it an attractive alternative to traditional methods, and its potential applications have inspired research and commercial opportunities [1, 12]. This protocol demonstrates a workflow for rapid plant diagnostics and marker-assisted breeding using LAMP genotyping.

2 Materials

2.1 Sample Preparation

1. Bunsen burner.
2. Calibrated p1, p100, p200 multichannel pipettes and p1, p200, p1000 single-channel pipettes.
3. 100–1000 μL nuclease-free aerosol filtered pipette tips.
4. 20–300 μL nuclease-free aerosol filtered pipette tips.
5. 1–100 μL nuclease-free aerosol filtered pipette tips.
6. 0.1–20 μL nuclease-free aerosol filtered pipette tips.
7. Stainless steel dissecting forceps.
8. 250 mL beaker.
9. Optional: plate piercer.
10. 1.5 mL centrifuge tubes.
11. Gloves.
12. 100% ethanol.
13. Nuclease-free water.
14. Extraction buffer solution: 0.25 M Tris–HCl, pH 8, 0.25% Triton-X.
15. Salt buffer solution: 5 M NaCl.
16. FTA plant saver cards.
17. Unicore punch kit, 2 mm.

2.2 LAMP Setup

1. Standard thermocycler.
2. Mini plate spinner.
3. Gloves.
4. Clean designated space.
5. Clear 96-well plate or similar.
6. Sterile loading trough.
7. B1, B2, B3, F1, F2, F3, BIP, FIP, LB oligonucleotides for LAMP analysis (*see* **Note 1**).
8. Colorimetric LAMP 2× Master Mix (DNA and RNA) (New England Biolabs) (*see* **Note 2**).
9. Genotype-specific positive DNA for the positive control template.

3 Methods

3.1 Sample Preparation

1. Prepare the workstation on the lab bench by wiping down the surface with alcohol where the preparation will be performed.
2. Prepare a clean PCR setup with clean 1.5 mL microcentrifuge tubes for sample preparation in an easily accessible area.
3. Sterilize forceps by dipping them in alcohol and flame sterilizing them or aseptically wiping excess alcohol until all alcohol has evaporated.
4. Remove a small sampling of raw tissue pieces or imprinted FTA card (~2 mm) with flame-sterilized forceps or alcohol-sterilized QIAGEN 2 mm UniCore Punch tool, taking care not to touch the sample directly with gloves.
5. Carefully place the sample into a 1.5 mL microcentrifuge tube at the bottom of the well so that it does not contact the top of the tube.
6. Repeat **steps 3–5** with remaining lots/replicates until all samples are processed.
7. Fill the sterile liquid basin/trough with Extraction Buffer using a dedicated single-channel or multichannel 30–300 μL multichannel pipettor loaded with clean tips to draw and dispense 200 μL of extraction buffer into each tube.
8. Repeat for all remaining samples and replace tips each time—cap tubes carefully in between steps.
9. Spin tubes down in microcentrifuge for 60 s to ensure all material is fully immersed in the extraction buffer.
10. Load the tube on a thermal incubation block and set to run 10 min "Boil" with lids sealed with tube lock to close.

11. ~95–100 °C 10 min 1× (*see* **Note 3**)
12. Following Boil, add 8 μL 5 M NaCl and vortex.
13. Add 400 μL of100% ethanol and invert several times.
14. Place tubes in −80 °C for 15 min or −20 °C overnight to precipitate DNA.
15. Following ultra-cold precipitation, mark the position where the pellet will form on the tube and centrifuge 1.5 mL tubes for 30 min at 4 °C.
16. Remove tubes carefully, decant the supernatant, and retain the pellet.
17. Optional: Perform a 70% ethanol wash to remove additional residual waste.
18. Following wash, centrifuge 1.5 mL tubes at 12,000 × *g* or greater for 30–60 min at 4 °C.
19. Remove tubes carefully, decant supernatant and retain the pellet (*see* **Note 4**).
20. Allow pellets to dry for 5–10 min at room temperature, then reconstitute the dried pellets in 50 μL nuclease-free water.

3.2 LAMP Setup

1. Prepare a clean LAMP PCR setup with a clean 96-well plate for sample preparation in an easily accessible area.
2. Prepare 25 μL final reactions with colorimetric LAMP Master Mix according to the following:

Reagent	Volume for 1 reaction (μL)
Colorimetric LAMP 2× Master Mix	12.5
LAMP Assay Primer Mix (20×)	2.5 (*see* **Note 5**)
Nuclease-Free Water	9.25

3. In the dedicated PCR space, use a 0.1–10 μL multichannel pipette, load 0.1–20 μL filtered pipette tips, and pipette 24 μL of the reaction mix to each well of the plate, respectively, for each sample.
4. Change tips and prepare to load samples, use fresh tips for each new sample.
5. Template reactions with purified total RNA or genomic DNA, using a 0.1–10 μL multichannel pipette or single-channel pipette, draw and dispense 1 μL of kit (1–10 ng/μL)* from purification kit, be sure tips are below surface of mix before dispensing, mix by pipetting up and down several time, then, eject tips after final mix.

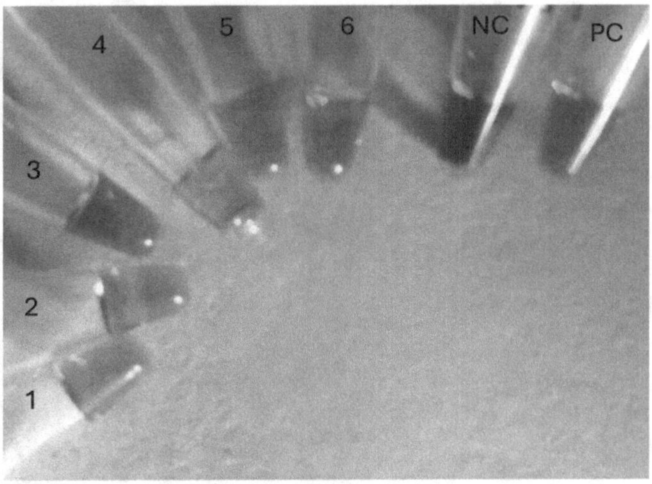

Fig. 1 Eight Reaction tubes at pre-isothermal incubation. 1–6 are test samples, 7 is a negative control, and 8 is a positive control

6. After each sample well has been filled, obtain the control standards and aliquot 1 μL of genotype specific positive control template, negative control template, and no template control into their own appropriate wells (Fig. 1).
7. Seal plate thoroughly with AlumaSeal sealing foil.
8. Spin plates for 10 s.
9. The plate is now ready for isothermal amplification and result visualization.
10. Place the LAMP plate in the thermocycler instrument (Fig. 2) and select and run the LAMP protocol:

    65 °C for 45 min 1× (@ 30 min begin to check results then again every 10 min (*see* **Note 6**).
11. Once the run is complete, remove the plate from the thermocycler and allow it to acclimate to room temperature.
12. Using a camera capture visual identification of LAMP results (Fig. 3).

4 Notes

1. Oligonucleotide design for LAMP analysis must be carried out carefully and specifically. Please utilize the design tool at Use Eikens Primer Explorer LAMP Assay Design Tool (link: https://primerexplorer.jp/e/) for design assistance and thermodynamic assessment. Optimization can enhance amplification efficiency and specificity in the early development of LAMP primer sets. It is essential to have B3, F3, FIP, BIP,

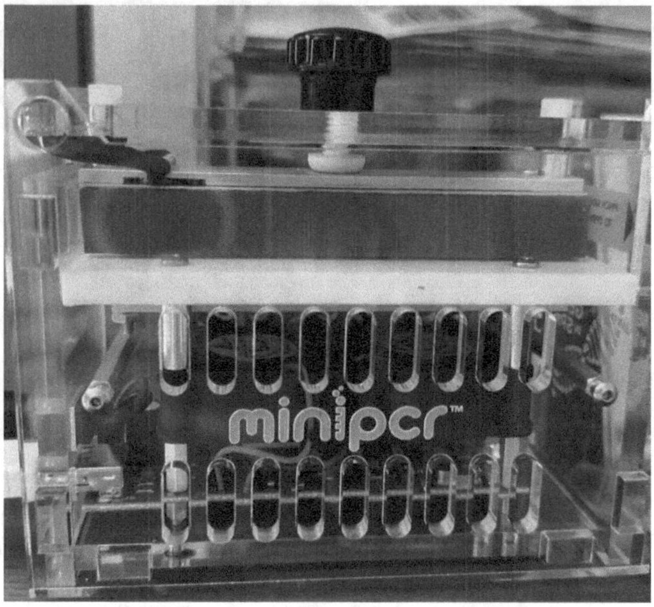

Fig. 2 Eight-well mini PCR thermocycler

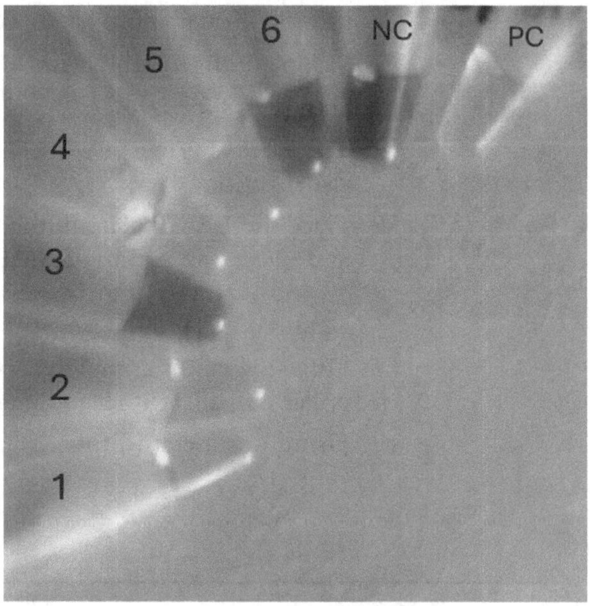

Fig. 3 Eight reaction tubes at post-isothermal incubation. 1–6 are test samples, 7 is a negative control, and 8 is a positive control

and LB primer sets, and additionally, adding B1, B2, F1, and F2 can enhance amplicon scaffolding.

2. This protocol utilizes the colorimetric LAMP Master Mix from NEB for visual detection; however, other LAMP/isothermal master mixes with alternative detection means can be explored.
3. Subheading 3.1, **steps 5–11** can be carried out in a 96-well plate for higher sample throughput. However, to complete DNA purification using sodium ethanol precipitation, high-speed centrifugation is required (>12,000 × *g*), and samples can be transferred to 1.5 microcentrifuge tubes at this point.
4. It is important to ensure that no residual ethanol remains before reconstitution, as it can inhibit LAMP downstream. Make sure enough time has been allowed to dry the pellet. As you marked the expected position of the pellet, you will be able to track it through the steps. However, sometimes it can be invisible to the naked eye, so be sure to wash all sides of the tube when you reconstitute the DNA.
5. Formulation of the 20× primer mix should follow this schema:

   LAMP Oligonucleotide: Volume of 100 μM Stock: Final 20× Conc. (μM)

   F3: 2 μL
   B3: 2 μL
   FIP: 16 μL
   BIP: 16 μL
   F2: 2 μL
   F1c: 2 μL
   B2. 2 μL
   B1c: 2 μL
   LB: 4 μL
   Nuclease-free water: 52 μL
6. Generally, visual identification results of color change indicating a positive reaction can begin to be observed after 30 min; however, additional incubation at 10-min intervals at 65 °C can be performed for low-level input or difficult to amplify targets.

Conflict of Interest

The authors have no conflicts of interest to declare that are relevant to the content of this chapter.

References

1. Fernandez IMA, Parungao M, Hollin J, Selimotic B, Farrar G, Seyler T, Anand A, Ahmad R (2023) A novel, precise and high-throughput technology for viroid detection in cannabis (MFDetect(TM)). Viruses 15(7): 1487. https://doi.org/10.3390/v15071487
2. Hasegawa Y, Ueno S, Wei FJ, Matsumoto A, Ujino-Ihara T, Uchiyama K, Moriguchi Y, Kasahara M, Fujino T, Shigenobu S, Yamaguchi K, Bino T, Hakamata T (2020) Development of diagnostic PCR and LAMP markers for MALE STERILITY 1 (MS1) in Cryptomeria japonica D. Don. BMC Res

Notes 13(1):457. https://doi.org/10.1186/s13104-020-05296-8

3. Liu X, Zhang C, Zhao M, Liu K, Li H, Li N, Gao L, Yang X, Ma T, Zhu J, Hui W, Hua K, Cui Y (2018) A direct isothermal amplification system adapted for rapid SNP genotyping of multifarious sample types. Biosens Bioelectron 115:70–76. https://doi.org/10.1016/j.bios.2018.05.021
4. Soroka M, Wasowicz B, Rymaszewska A (2021) Loop-mediated isothermal amplification (LAMP): the better sibling of PCR? Cells 10(8):1931. https://doi.org/10.3390/cells10081931
5. Vanhomwegen J, Kwasiborski A, Diop A, Boizeau L, Hoinard D, Vray M, Bercion R, Ndiaye B, Dublineau A, Michiyuki S, Manuguerra JC, Sauvage V, Candotti D, Seck A, Laperche S, Shimakawa Y (2021) Development and clinical validation of loop-mediated isothermal amplification (LAMP) assay to diagnose high HBV DNA levels in resource-limited settings. Clin Microbiol Infect 27(12):1858.e1859–1858.e1815. https://doi.org/10.1016/j.cmi.2021.03.014
6. Torres A, Pauli C, Givens R, Argyris J, Allen K, Monfort A, Gaudino RJ (2022) High-throughput methods to identify male Cannabis sativa using various genotyping methods. J Cannabis Res 4(1):57. https://doi.org/10.1186/s42238-022-00164-7
7. Dassanayake RP, Clawson ML, Tatum FM, Briggs RE, Kaplan BS, Casas E (2023) Differential identification of Mannheimia haemolytica genotypes 1 and 2 using colorimetric loop-mediated isothermal amplification. BMC Res Notes 16(1):4. https://doi.org/10.1186/s13104-023-06272-8
8. Zhou D, Guo J, Xu L, Gao S, Lin Q, Wu Q, Wu L, Que Y (2014) Establishment and application of a loop-mediated isothermal amplification (LAMP) system for detection of cry1Ac transgenic sugarcane. Sci Rep 4:4912. https://doi.org/10.1038/srep04912
9. Moehling TJ, Choi G, Dugan LC, Salit M, Meagher RJ (2021) LAMP diagnostics at the point-of-care: emerging trends and perspectives for the developer community. Expert Rev Mol Diagn 21(1):43–61. https://doi.org/10.1080/14737159.2021.1873769
10. Crego-Vicente B, Del Olmo MD, Muro A, Fernandez-Soto P (2024) Multiplexing LAMP assays: a methodological review and diagnostic application. Int J Mol Sci 25(12):6374. https://doi.org/10.3390/ijms25126374
11. Kim SH, Lee SY, Kim U, Oh SW (2023) Diverse methods of reducing and confirming false-positive results of loop-mediated isothermal amplification assays: a review. Anal Chim Acta 1280:341693. https://doi.org/10.1016/j.aca.2023.341693
12. Lee D, La Mura M, Allnutt T, Powell W, Greenland A (2009) Isothermal amplification of genetically modified DNA sequences directly from plant tissues lowers the barriers to high-throughput and field-based genotyping. J Agric Food Chem 57(20):9400–9402. https://doi.org/10.1021/jf902455j

Chapter 6

RPA Workflow for Sex Determination Genotyping in a Dioecious Plant

Anthony Torres and Reginald Gaudino

Abstract

Field-based techniques for genotyping and identification are facilitated by isothermal amplification and detection technology. This chapter focuses on developing a rapid and field-deployable method for early sex identification genotyping in plants. Leveraging the Y-chromosome-specific regions or gene segments, we designed and evaluated an alternative isothermal amplification assay through recombinase polymerase amplification (RPA) to detect male-specific DNA sequences and perform sex determination genotyping.

Key words DNA, Amplicon, Isothermal, Primers, Allele

1 Introduction

Preliminary analysis with the TwistDx RPA system demonstrated successful differentiation of male and female plants, offering a promising solution for early sex genotyping and resource optimization in marker-assisted breeding in a dioecious plant model [1–3]. RPA employs a unique combination of enzymes, including recombinase, single-stranded DNA binding protein, and strand-displacing polymerase, to enable DNA amplification at a constant temperature [4]. This eliminates the need for thermal cycling, making it suitable for on-site testing [5]. In other crops, RPA could enable the early detection of disease-resistance genes, facilitating the selection of resilient plants and reducing reliance on pesticides [6, 7].

Traditional sex identification methods rely on observing phenotypic traits, which often manifest weeks into the flowering stage, leading to wasted resources on male plants. Modern agricultural biotechnology molecular techniques, such as PCR-based assays targeting Y-chromosome-specific sequences for male genotype identification, provide a more efficient approach [1, 3]. However, these methods often require laboratory equipment and expertise,

Anthony Torres (ed.), *Genotyping: Methods and Protocols*, Methods in Molecular Biology, vol. 2943,
https://doi.org/10.1007/978-1-0716-4642-7_6,

limiting their field applicability. Its simplicity, speed, and portability make it a reliable tool for plant breeders and researchers seeking to accelerate crop improvement efforts and tackle challenges associated with climate change, pests, and diseases [5, 8, 9].

While further research and development are required to fully harness the potential of RPA, its ability to bridge the gap between laboratory-based genotyping and field applications marks a significant step forward in agricultural biotechnology [10, 11].

In this chapter, we explored the use of Recombinase Polymerase Amplification (RPA), an isothermal amplification method, for developing a rapid and field-deployable assay for early sex identification in *C. sativa* [1]. RPA stands out from traditional PCR methods due to its significantly faster reaction times, simplified workflows, and the potential for visual detection in a simplified workflow.

2 Materials

2.1 Sample Preparation

1. Bunsen burner.
2. Calibrated p1, p100, p200 multichannel pipettes and p1, p200, p1000 single-channel pipettes.
3. 100–1000 μL nuclease-free aerosol filtered pipette tips.
4. 20–300 μL nuclease-free aerosol filtered pipette tips.
5. 1–100 μL nuclease-free aerosol filtered pipette tips.
6. 0.1–20 μL nuclease-free aerosol filtered pipette tips.
7. Stainless steel dissecting forceps.
8. 250 mL beaker.
9. Optional: plate piercer.
10. 1.5 mL centrifuge tubes.
11. Gloves.
12. 100% ethanol.
13. Nuclease-free water.
14. Extraction buffer solution: 0.25 M Tris–HCl, pH 8, 0.25% Triton-X.
15. Salt buffer solution: 5 M NaCl.
16. FTA plant saver cards.
17. Unicore punch kit, 2 mm.

2.2 RPA Reaction Setup

1. Standard thermocycler.
2. Mini plate spinner.
3. Gloves.

4. Clean designated space.
5. Clear 96-well plate or similar.
6. Sterile loading trough.
7. Custom Biotin Labeled fwd and FITC labeled rev oligonucleotides (*see* **Note 1**).
8. TwistAmp™ Basic Amp Reaction Kit Product Number: INTA-BAS (TwistDX) (*see* **Note 2**).
9. Primer free rehydration buffer.
10. Lyophilized reaction pellet.
11. Genotype-specific positive DNA for the positive control template.

2.3 RPA Genotyping Detection

1. Milenia Genline HybriDetect strips (Milenia Genline, Germany) (*see* **Note 2**).
2. USTAR disposable nucleic acid detection device (Type 3) (USTAR, China) (*see* **Note 2**).

3 Methods

3.1 Sample Preparation

1. Prepare the workstation on the lab bench by wiping down the surface with alcohol where the preparation will be performed.
2. Prepare a clean PCR setup with clean 1.5 mL microcentrifuge tubes for sample preparation in an easily accessible area.
3. Sterilize forceps by dipping them in alcohol and flame sterilizing them or aseptically wiping excess alcohol until all alcohol has evaporated.
4. Remove a small sampling of raw tissue pieces or imprinted FTA card (~4–8 mm) with flame-sterilized forceps or alcohol-sterilized QIAGEN 2 mm UniCore Punch tool, taking care not to touch the sample directly with gloves. (*see* **Note 3**).
5. Carefully place the sample into a 1.5 mL microcentrifuge tube at the bottom of the tube so that it does not contact the top of the tube.
6. Repeat **steps 3.1.3–3.1.5** with remaining lots/replicates until all samples are processed.
7. Fill the sterile liquid basin/trough with Extraction Buffer using a dedicated single-channel or multichannel 30–300 μL multichannel pipettor loaded with clean tips to draw and dispense 200 μL of extraction buffer into each tube.
8. Repeat for all remaining samples and replace tips each time—cap tubes carefully in between steps.

9. Spin tubes down in microcentrifuge for 60 s to ensure all material is fully immersed in the extraction buffer.
10. Load the tube on a thermal incubation block and set to run 10 min "Boil" with lids sealed with tube lock to close.
11. ~95–100 °C 10 min 1× (*see* **Note 4**).
12. Following Boil, add 8 μL 5 M NaCl and vortex.
13. Add 400 μL of 100% ethanol and invert several times.
14. Place the tubes in −80 °C for 15 min or −20 °C overnight to precipitate DNA.
15. Following ultra-cold precipitation, mark the position where the pellet will form on the tube and centrifuge 1.5 mL tubes for 30 min at 4 °C.
16. Remove tubes carefully, decant the supernatant, and retain the pellet.
17. Optional: Perform a 70% ethanol wash to remove additional residual waste.
18. Following wash, centrifuge 1.5 mL tubes at >10,000 × *g* or greater for 30–60 min at 4 °C.
19. Remove tubes carefully, decant supernatant and retain the pellet (*see* **Note 5**).
20. Allow pellets to dry for 5–10 min at room temperature, then reconstitute the dried pellets in 50 μL nuclease-free water.

3.2 RPA Reaction Setup

1. Prepare a clean RPA PCR setup with a clean 8-tube strip for sample preparation in an easily accessible area.
2. Prepare 50 μL final reactions with TwistAmp Reaction Pellet and Custom Genotyping Primer Mix (*see* **Note 1**) according to the following:

Reagent	Volume for 1 reaction (μL)
Primer Free Rehydration Buffer	29.5
Biotin Labeled Fwd Primer (10 μM)	2.4
FITC Labeled Rev Primer (10 μM)	2.4
Nuclease-free water	3
Template (Each tube)	10.2

3. In the dedicated PCR space, use a 0.1–10 μL multichannel pipette, load 0.1–20 μL filtered pipette tips, and pipette 37.5 μL of the reaction mix to each lyophilized tube for each sample.

4. Change tips and prepare to load samples, use fresh tips for each new sample.
   (a) For all reactions with sample preparation, using a 0.1–10 μL multichannel pipette or single-channel pipette, draw and dispense 10.2 μL of reconstituted DNA input.
   (b) For all reactions with purified genomic DNA, using a 0.1–10 μL multichannel pipette or single-channel pipette, draw and dispense 10.2 μL of kit (1–10 ng/μL)* purified genomic DNA from DNEasy or similar, be sure tips are below surface of mix before dispensing, mix by pipetting up and down several time, then, eject tips after final mix.
5. After each sample tube has been filled, obtain the control standards and aliquot 10.2 μL of positive control template, negative control template, and no template control into their own appropriate tubes.
6. Add 2.5 μL of 250 mM Magnesium Acetate (MgOAc) (TwistAmp) to each sample to start the reaction. This brings the final reaction volume to 50 μL.
7. Seal tubes thoroughly with a capped seal.
8. Spin tubes for 10 s.
9. The tubes are now ready for isothermal amplification.
10. Place the RPA tube in the thermocycler instrument and select and run the RPA protocol:
    (a) 39 °C for 20 min 1×. Alternatively, run at room temperature for 30 min (*see* **Note 6**).
11. After isothermal incubation, pull the tube out of the thermocycler and allow it to acclimate to room temperature.

3.3 RPA Genotyping Detection

1. For genotyping results visualization, the Milenia Genline HybriDetect strips (Milenia Genline, Germany) loaded into a USTAR disposable nucleic acid detection device (Type 3) (USTAR, China) are used.
2. Attach the finished tube to the tube holder of the device cartridge and fold the cartridge lid closed to lock the tube into place (*see* **Note 7**).
3. Insert the device cartridge into the Type 3 detection device and push the handle to a locked position (USTAR, China). This step pierces the tube.
4. Wait 15–30 min for lateral flow to carry out the reaction across the test strip (*see* **Note 8**).
5. Successful amplification of the target genotype will result in two bands on the test strip indicative of positive detection of the genotype-specific target amplicon labeled with biotin and FITC and conjugation with the test strip, as well as a reaction

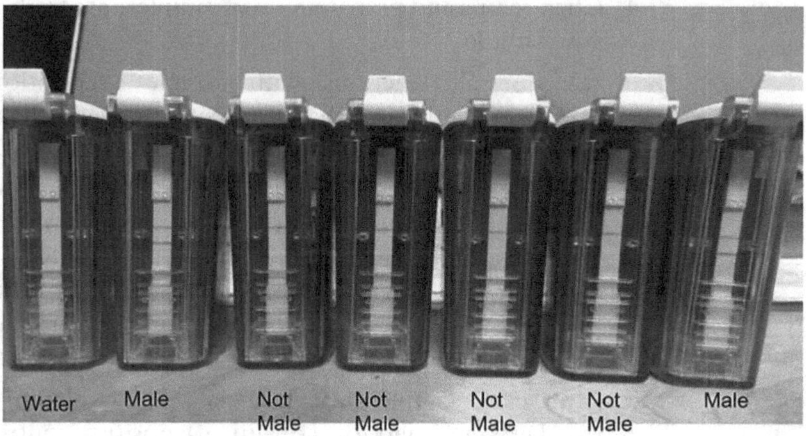

Fig. 1 TwistDX assay results of male and not male gender determination of *C. sativa* plant samples using a USTAR Type 3 detection device [1]

control band indicating a successful reaction. If only a single control band is present, the sample is absent of the target genotype (*see* **Note 9**).

6. Using a camera, capture visual identification of RPA results (Fig. 1).

4 Notes

1. Oligonucleotides must be well designed. One primer pair will be used per detection assay. The primer pair must be unique and genotype-specific to produce a result detecting a specific genotype, such as Y chromosome detection of a male genotype of a dioecious plant. The 5′ end of the oligonucleotides must be labeled accordingly. The forward primer should be modified and labeled with a 5′ tagBiotin molecule, and the reverse primer should be labeled with a Fluorescein (FITC) molecule.
2. The TwistAmp™ Basic Amp Reaction Kit is produced by TwistDX and can be utilized with any custom-designed primer set. While other commercial RPA kits are available, this protocol utilizes this product along with the DNA detection on lateral flow strip from Milenia Genline HybriDetect strips and Detection by USTAR disposable nucleic acid detection device (Type 3).
3. All samples must be labeled with a unique identification number, and an alphanumeric well position number must be assigned for all future labeling. Additionally, if FTA imprinted spots are light, use greater sampling size (~8 mm) in same extraction volume (100 μL).

4. Subheading 3.1, **steps 5–11** can be carried out in a 96-well plate for higher sample throughput. However, to complete DNA purification using sodium ethanol precipitation, high-speed centrifugation is required (>12,000 × *g*), and samples can be transferred to 1.5 microcentrifuge tubes at this point.
5. It is important to ensure that no residual ethanol remains before reconstitution, as it can inhibit PCR downstream. Make sure enough time has been allowed to dry the pellet. As you marked the expected position of the pellet, you will be able to track it through the steps. However, sometimes it can be invisible to the naked eye, so wash all sides of the tube when you reconstitute the DNA.
6. Incubating on a thermocycler ensures stable isothermal temperature during incubation. If incubation is performed at room temperature, be sure to maintain a constant temperature. Room temperature incubation time can be extended for low-threshold detection.
7. Once the template has been added and RPA incubation occurs, keep the tubes closed throughout the process. The tubes will be inserted into the detection device, which will be capped so that the device can pierce the tube properly.
8. Lateral flow can be observed visually to ensure efficient fluidic transfer across the flow strip. Allow at least 20–30 min for full lateral flow.
9. Each lateral flow strip has a control detection along with the target detection strip. The control strip confirms that RPA occurred, and lateral flow was successful; the second band will be the target genotype detection result for the custom design RPA assay. The detection happens by immunoprecipitation reaction of detected target DNA.

Conflict of Interest

The authors have no conflicts of interest to declare that are relevant to the content of this chapter.

References

1. Torres A, Pauli C, Givens R, Argyris J, Allen K, Monfort A, Gaudino RJ (2022) High-throughput methods to identify male Cannabis sativa using various genotyping methods. J Cannabis Res 4(1):57. https://doi.org/10.1186/s42238-022-00164-7
2. Zhou J, Zhang A, Xu W, Cheng C, Niu F, Sun B, Cao L, Zhang J, Chu H (2024) Rapid detection of rice fragrance allele badh2-E7 by recombinant polymerase amplification (RPA). Biosci Methods. https://doi.org/10.5376/bm.2024.15.0001
3. Ávila-Hernández JG, Coreño-Alonso A, Pantoja-Alonso MA, Córdoba-Andrade FJ, Camas-Reyes A, Martínez-Antonio A (2024) Step-by-step development of a recombinase polymerase amplification (RPA) assay for sex

identification in papaya. Preprints. https://doi.org/10.20944/preprints202408.2108.v1

4. Tan M, Liao C, Liang L, Yi X, Zhou Z, Wei G (2022) Recent advances in recombinase polymerase amplification: principle, advantages, disadvantages and applications. Front Cell Infect Microbiol 12:1019071. https://doi.org/10.3389/fcimb.2022.1019071
5. Chen W-y, Yu J, Xu H, Lu X-y, Dai T-t, Tian Y-e, Shen D-y, Dou D-l (2021) Combining simplified DNA extraction technology and recombinase polymerase amplification assay for rapid and equipment-free detection of citrus pathogen Phytophthora parasitica. J Integr Agric 20(10):2696–2705. https://doi.org/10.1016/s2095-3119(20)63459-1
6. Clark KJ, Anchieta AG, da Silva MB, Kandel SL, Choi YJ, Martin FN, Correll JC, Van Denyze A, Brummer EC, Klosterman SJ (2022) Early detection of the spinach downy mildew pathogen in leaves by recombinase polymerase amplification. Plant Dis 106(7):1793–1802. https://doi.org/10.1094/PDIS-11-21-2398-RE
7. Higgins DS, Miles TD, Byrne JM, Hausbeck MK (2022) Optimizing molecular detection for the hop downy mildew pathogen Pseudoperonospora humuli in plant tissue. Phytopathology 112(11):2426–2439. https://doi.org/10.1094/PHYTO-01-22-0013-R
8. Lee D, La Mura M, Allnutt T, Powell W, Greenland A (2009) Isothermal amplification of genetically modified DNA sequences directly from plant tissues lowers the barriers to high-throughput and field-based genotyping. J Agric Food Chem 57(20):9400–9402. https://doi.org/10.1021/jf902455j
9. Wang TM, Yang JT (2019) Visual DNA diagnosis of Tomato yellow leaf curl virus with integrated recombinase polymerase amplification and a gold-nanoparticle probe. Sci Rep 9(1):15146. https://doi.org/10.1038/s41598-019-51650-7
10. Boluk G, Dobhal S, Crockford AB, Melzer M, Alvarez AM, Arif M (2020) Genome-informed recombinase polymerase amplification assay coupled with a lateral flow device for in-field detection of Dickeya species. Plant Dis 104(8):2217–2224. https://doi.org/10.1094/PDIS-09-19-1988-RE
11. Priti, Jangra S, Baranwal VK, Dietzgen RG, Ghosh A (2021) A rapid field-based assay using recombinase polymerase amplification for identification of Thrips palmi, a vector of tospoviruses. J Pest Sci (2004) 94(2):219–229. https://doi.org/10.1007/s10340-020-01284-w

Chapter 7

Sex Genotyping by Flow Cytometry

Naoki Uno

Abstract

A flow cytometry assay for sex genotyping at single cell resolution is described. This assay is based on the detection of a Y chromosome-specific gene by in situ PCR. The protocol requires the isolation of lymphocytes from human blood, their fixation, permeabilization, and mild protein digestion, and the amplification of a Y chromosome-specific gene *SRY*, which is present only in male cells, by in situ PCR with a universal energy transfer-labelled primer (UniPrimer). Male cells can be detected by flow cytometry because the *SRY* gene is amplified and fluorescently labelled solely in male cells.

Key words In situ PCR, Y chromosome-specific genes, Universal energy transfer-labelled primer, Flow cytometry

1 Introduction

Sex genotyping can be achieved simply by amplifying a Y chromosome-specific gene by PCR [1]. This protocol is intended for the detection of male cells at the single cell level by flow cytometry. Male cells can be detected by fluorescent microscopy if the Y chromosome-specific genes are labelled by fluorescent in situ hybridization (FISH). However, fluorescence signals generated from FISH-labelled genomic DNA are not strong enough to be detected by flow cytometry. This limitation can be avoided by amplifying target DNA using in situ PCR. However, the traditional in situ PCR technique, which is based on incorporation of digoxigenin-labelled dUTP during PCR [2], suffers from strong background noise, because digoxigenin-labelled dUTPs are readily incorporated into nonspecific PCR products even in the absence of primers [3]. This protocol is based on another in situ PCR technique that relies on PCR products labelled with a universal energy transfer-labelled primer (UniPrimer) (Fig. 1) [4]. I determined the optimal conditions that eliminate background noise, resulting in a signal-to-noise ratio that is high enough to detect a single-copy

Anthony Torres (ed.), *Genotyping: Methods and Protocols*, Methods in Molecular Biology, vol. 2943,
https://doi.org/10.1007/978-1-0716-4642-7_7,

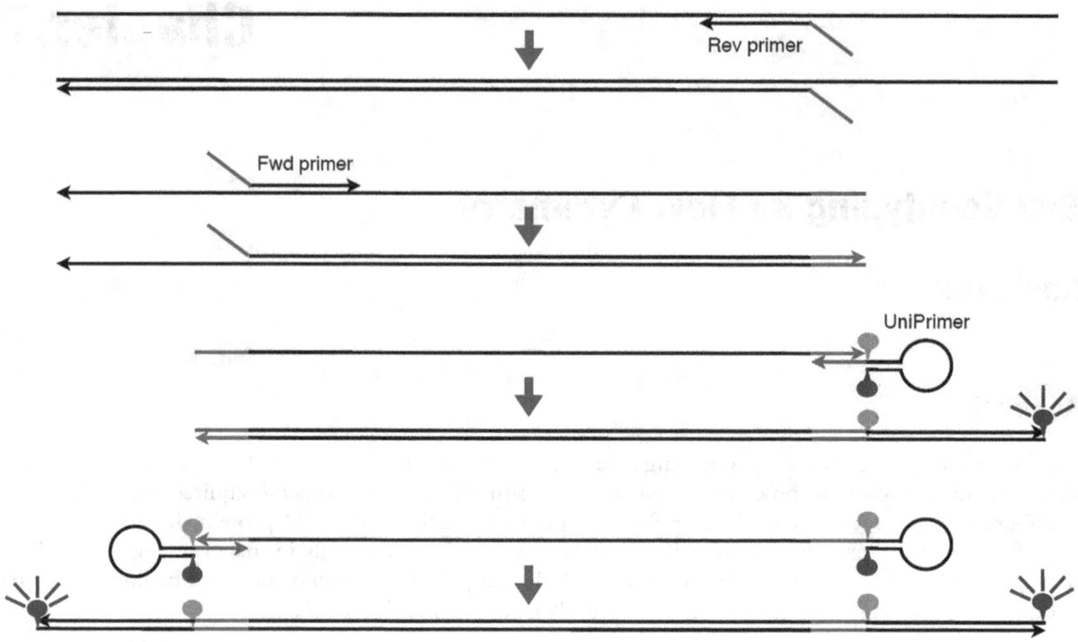

Fig. 1 Schematic of in situ PCR with the UniPrimer. Target-specific primers contain a tail sequence at their 5′ end (red) that is identical to the 3′ end sequence of the UniPrimer. The sequence complementary to the UniPrimer (blue) does not exist before PCR but is synthesized when target-specific primers are elongated during the initial stage of PCR. The UniPrimer then hybridizes to the 3′ ends of the elongated target-specific primers, resulting in the amplification of fluorescein-labelled products during subsequent cycles. Thus, the UniPrimer does not cause background noise in the absence of target DNA but generates strong signals when the target DNA is present

DNA target in human lymphocytes by flow cytometry [5]. Herein, I describe a protocol designed to detect human lymphocytes that contain the *SRY* gene, which is present as a single copy per male cell [6], by flow cytometry.

2 Materials

2.1 Isolation of Lymphocytes from Human Blood

1. Blood collection tubes containing either heparin or EDTA.
2. EasySep Direct Human Cell Isolation Kits (STEMCELL Technologies) (*see* **Note 1**).
3. EasySep magnet (STEMCELL Technologies) (*see* **Note 1**).
4. 14 mL round bottom polystyrene tubes (Corning, 352017).
5. 15 mL centrifuge tubes.
6. Serological pipettor.
7. 2 mL and 10 mL serological pipettes.
8. 1.5 mL tubes.
9. Centrifuge.

2.2 Fixation and Permeabilization of Isolated Lymphocytes

1. BD Cytofix/Cytoperm fixation and permeabilization solution (BD Biosciences, 554722).
2. 1× BD Perm/Wash buffer: Dilute 10× BD Perm/Wash buffer (BD Biosciences, 554723) with water. Store at 4 °C.
3. Cell counter.

2.3 Protease Digestion

1. Proteinase K stock solution (2 mg/mL): Dilute Proteinase K solution (Roche Diagnostics, 03 115 828 001) with water or phosphate-buffered saline (PBS). Store at 4 °C.
2. Proteinase K working solution (50 μg/mL): Dilute the Proteinase K stock solution with 1× BD Perm/Wash buffer. Prepare before use.
3. Heat block.

2.4 In Situ PCR

1. 0.2 mL individual PCR tubes.
2. Thermocycler.
3. Microfuge.
4. Oligonucleotide primers targeting *SRY* and *HBB*. The *SRY* gene is located only on the Y chromosome and therefore is present only in male cells, whereas the *HBB* gene is located on an autosome and is therefore present in both female and male cells. All primers should contain a tail sequence (underlined) that is identical to the 3′ end of the UniPrimer.

   SRY forward: 5′ ACTGAACCTGACCGTACAGCTGGGATACCAGTGGAAAA 3′.

   SRY reverse: 5′ ACTGAACCTGACCGTACATTCCGACGAGGTCGATACTT 3′.

   HBB forward: 5′ ACTGAACCTGACCGTACAGTGCATCTGACTCCTGAGGAGA 3′.

   HBB reverse: 5′ ACTGAACCTGACCGTACACCTTGATACCAACCTGCCCAG 3′.
5. Amplifluor UniPrimer II (Merck, S7905).
6. TaKaRa Ex Taq HS (Takara Bio, RR006) (*see* **Note 2**).
7. Ex Taq Buffer (Takara Bio).
8. dNTPs mixture.

2.5 Flow Cytometry Analysis

1. 5 mL round bottom polystyrene tube (Corning, 352052).
2. PBS.
3. BD flow cytometry analyzer.
4. BD flow cytometry sheath fluid.

3 Methods

3.1 Isolation of Lymphocytes from Human Blood

This protocol is validated for use with human lymphocytes and neutrophils that have been freshly separated from whole blood but is NOT validated for use with other cells such as cell lines cultured in vitro. Note that the assay is heavily dependent on the cell type. If another type of cell is used, the optimal conditions for that cell type will need to be determined. The optimal conditions for other cells may vary considerably and may even be impossible to determine.

1. Collect peripheral blood in a tube containing either heparin or EDTA.
2. Separate lymphocytes from whole blood using an EasySep Direct Human Cell Isolation Kit according to the manufacturer's instruction (*see* **Note 1**).
3. Count cells and transfer the appropriate number of cells into a 1.5 mL tube (*see* **Note 3**).

3.2 Fixation and Permeabilization of Isolated Cells

The use of BD buffers described in Subheading 2.2 is recommended. The protocol is not validated for use with other fixatives or wash buffers.

1. Centrifuge the tube at 3000 × *g* for 5 min and remove supernatant.
2. Fix cells by resuspending the cells in 100 μL of BD Cytofix/Cytoperm fixation and permeabilization solution.
3. Incubate at room temperature for 30 min (*see* **Note 4**).
4. Add 1 mL of 1× BD Perm/Wash buffer.
5. Leave the tube overnight at 4 °C (*see* **Note 5**).

3.3 Preparation of PCR Reaction Solution

1. Assemble the PCR reaction in a 1.5 mL tube as follows (*see* **Note 6**):

   1× EX Taq Buffer.

   0.2 mM dNTPs.

   0.1 μM forward and reverse primers targeting either *SRY* or *HBB* (*see* **Note 7**).

   0.5 μM Amplifluor UniPrimer II.

   0.05 U/μL TaKaRa EX Taq HS.

   Water.

2. Keep the reaction mixture on ice until **step 3** in Subheading 3.5.

3.4 Protease Digestion

Protease digestion is necessary to prevent background noise. This step should be optimized and executed carefully.

1. Prepare Proteinase K working solution (50 μg/mL) as described in Subheading 2.3 (*see* **Note 8**).
2. Centrifuge the tube containing cells at 3000 × *g* for 5 min.
3. Remove supernatant.
4. Resuspend the cells in 100 μL Proteinase K working solution.
5. Incubate at 40 °C for 15 min (*see* **Note 8**).
6. Add 1 mL of 1× BD Perm/Wash buffer, centrifuge at 3000 × *g* for 5 min, and remove supernatant.
7. Resuspend the cells in 1× BD Perm/Wash buffer and transfer the cells to individual PCR tubes so that 1 × 10^5 cells are present in a tube (*see* **Note 9**).

3.5 In Situ PCR

1. Centrifuge PCR tubes for 3 min in a microfuge.
2. Remove as much supernatant as possible using a pipette. Be careful not to disturb the pelleted cells (*see* **Note 10**).
3. Resuspend the cells in the PCR mixture prepared in Subheading 3.3. Use 30 μL of the PCR reaction mixture per tube (*see* **Note 11**).
4. Run the PCR program as shown below (*see* **Note 12**).

   Initial denaturation at 95 °C for 10 min.

   50 cycles of denaturation at 95 °C for 10 s, annealing at 63 °C for 20 s, extension at 72 °C for 40 s.

   Final extension at 72 °C for 5 min.

   Hold at 4 °C.

3.6 Flow Cytometry Analysis

1. Add 200–300 μL of PBS to a 5 mL round bottom polystyrene tube.
2. Add the PCR reaction mixture directly to the tube. No washing step is needed.
3. Startup the BD flow cytometry system according to the manufacturer's instruction.
4. Create an FSC/SSC dot plot, a FITC histogram, and statistics in a worksheet.
5. Load a sample tube and acquire data.
6. Gate a lymphocyte population in the FSC/SSC dot plot if needed.
7. Draw a histogram marker in the FITC histogram from the right edge of the population of the sample without polymerase and then assess positive events in the histogram generated from the sample with polymerase for the same target (*see* **Note 13**).

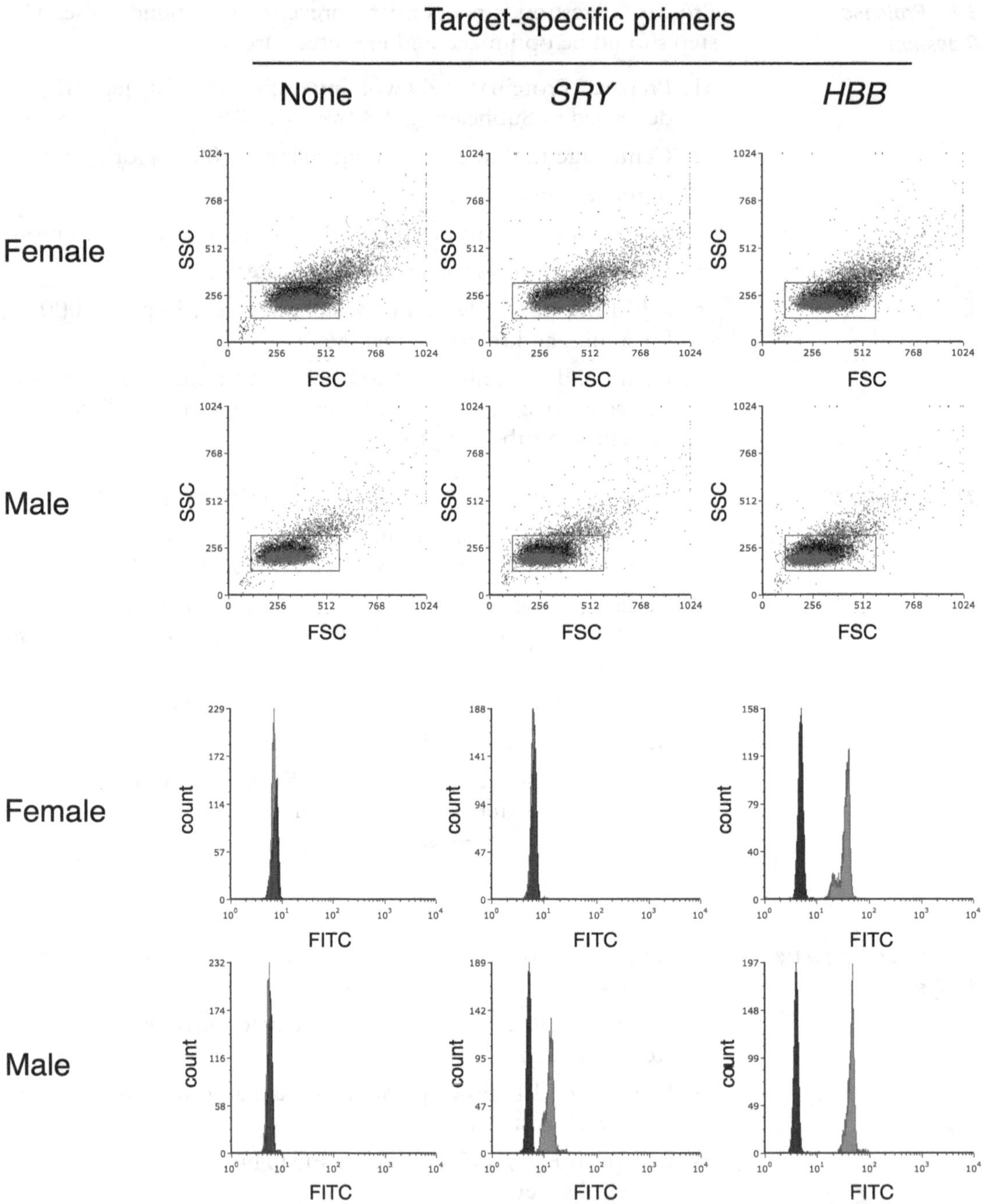

Fig. 2 In situ PCR was performed in the absence (black) and presence (red) of polymerase using CD4$^+$ T cells separated from whole blood of healthy female and male subjects. Histograms were created from populations gated in FSC/SSC dot plots. When *SRY*-specific primers are used, the histogram shows that populations in the absence and presence of polymerase totally overlap in a female subject but are completely separate in a male subject, indicating that the *SRY* gene is absent from all female cells but present in all male cells

8. Generate an overlaid histogram for each target. An example of the results that can be obtained using this protocol is shown in Fig. 2 (*see* **Notes 14** and **15**).

4 Notes

1. EasySep Direct Human Cell isolation kits are useful and efficient for the isolation of lymphocytes from whole blood. However, other methods may also be used as long as the human lymphocytes are separated properly from whole blood.
2. Another DNA polymerase may also be used. However, a hot start polymerase should be used to reduce as much as possible nonspecific amplification.
3. For example, 1–2 × 10^6 cells are transferred to a tube.
4. The incubation time affects the results and may be adjusted to between 5 and 30 min.
5. It is not necessary to leave cells overnight at 4 °C. It is possible to proceed directly to the next step. However, the overnight pause at 4 °C helps improve the reproducibility of results.
6. It is necessary to prepare control samples. Prepare samples without DNA polymerase for each target. Prepare samples in the absence of target-specific primers but in the presence of the UniPrimer to verify that the UniPrimer does not cause background noise. Prepare female and male lymphocytes as negative and positive controls, respectively.
7. Prepare PCR reaction mixtures for each target separately. Multiplex assays are not possible unless another UniPrimer containing a distinct 3′ end sequence and another fluorophore is created.
8. The optimal concentration of Proteinase K ranges from 100 to 50 mg/mL, and the optimum temperature ranges from 35 to 55 °C.
9. It is not possible to count cells accurately after Proteinase K digestion. Therefore, the number of cells should be estimated from the one counted in **step 3** in Subheading 3.1.
10. Pelleted cells are hardly visible.
11. The concentration of cells in the reaction affects results. It should be no greater than 2 × 10^5 cells/30 μL of reaction.
12. PCR products can be analysed by agarose gel electrophoresis after in situ PCR to confirm the generation of specific products. It is not necessary to lyse cells in this case, because amplification products move out through the cell membrane and therefore are present in the reaction solution.
13. If positive events are detected in the absence of target-specific primers, experimental conditions such as protease digestion and fixation must be optimized.

14. When *SRY*-specific primers are used, the overlaid histogram should show that cell populations in the absence and presence of polymerase are clearly separated in males. If they overlap, experimental conditions such as protease digestion and fixation should be optimized.
15. If a sample contains both female and male cells, the male fraction can be estimated but will not be accurate.

References

1. Papaioannou VE, Behringer RR (2024) Sex genotyping mice by polymerase chain reaction. Cold Spring Harb Protoc 2024(1):108062. https://doi.org/10.1101/pdb.prot108062
2. Nuovo GJ (1995) In situ PCR: protocols and applications. PCR Methods Appl 4(4):S151–S167. https://doi.org/10.1101/gr.4.4.s151
3. Nuovo GJ, MacConnell P, Gallery F (1994) Analysis of nonspecific DNA synthesis during in situ PCR and solution-phase PCR. PCR Methods Appl 4(2):89–96. https://doi.org/10.1101/gr.4.2.89
4. Nuovo GJ, Hohman RJ, Nardone GA, Nazarenko IA (1999) In situ amplification using universal energy transfer-labelled primers. J Histochem Cytochem 47(3):273–280. https://doi.org/10.1177/002215549904700301
5. Uno N, Kaku N, Morinaga Y, Hasegawa H, Yanagihara K (2020) Flow cytometry assay for the detection of single-copy DNA in human lymphocytes. Nucleic Acids Res 48(15):e86. https://doi.org/10.1093/nar/gkaa515
6. Bachtrog D, Charlesworth B (2001) Towards a complete sequence of the human Y chromosome. Genome Biol 2(5):REVIEWS1016. https://doi.org/10.1186/gb-2001-2-5-reviews1016

Chapter 8

High-Throughput Genotyping by CD-PCR

Yoichi Nakanishi and Shunta Taga

Abstract

The process of extracting genomic DNA from biological samples is a bottleneck in PCR genotyping. Here, we describe a speedy genotyping method that combines simple genomic DNA extraction using hot alkaline/SDS and direct use of the extract in an SDS-tolerant cyclodextrin-PCR (CD-PCR).

Key words Genotyping, SDS, Cyclodextrin, Cyclodextrin-PCR (CD-PCR), Arabidopsis, Yeast, High throughput, T7 endonuclease I assay

1 Introduction

Genotyping gene-labeled materials is one of the most widely used methods in all fields, from basic biology research to applied science, and its demand is increasing today. For example, in basic plant research, molecular genetic tools such as T-DNA tag deletion lines, modified endogenous genes using genome editing techniques, and reporter genes are combined to obtain new findings, and researchers are increasingly tasked with genotyping to check them. The introduction of PCR has dramatically simplified the gene determination process; however, preparing genomic DNA (gDNA) from test samples is still time-consuming.

Here, we demonstrate a straightforward, reproducible, time- and cost-saving genotyping method [1]. The preparation of gDNA by hot alkaline/SDS extraction from plants and yeast is effortless; alkali weakens the cell wall, peels off DNA-binding proteins; the detergent SDS destroys plasma and nuclear membranes, denatures intracellular proteins, and increases the efficiency of DNA extraction. Heating also increases extraction efficiency. Although the gDNA contains SDS, which denatures DNA polymerase and inhibits PCR, the inhibitory effect of SDS can be overcome by adding cyclodextrin (CD). This type of cyclic oligosaccharide has an inclusion ability of SDS [2, 3]. Also, the introduction of small amounts

Anthony Torres (ed.), *Genotyping: Methods and Protocols*, Methods in Molecular Biology, vol. 2943,
https://doi.org/10.1007/978-1-0716-4642-7_8,

of alkali does not affect PCR due to the buffer. The CD-PCR products can be analyzed directly using agarose gel electrophoresis. In addition, the PCR product can be subjected to post-PCR processes such as restriction enzyme reactions and T7 endonuclease I assay [4]. We also show guidelines on how to design PCR for the high-throughput experiment.

2 Materials

Prepare all solutions in the PCR composition free of genetic contamination (*see* **Note 1**). Use "PCR-specific" reagents and disposable or thoroughly cleaned equipment to prepare solutions (*see* **Note 2**). Use freshly prepared ultrapure water or commercially available ultrapure water.

2.1 Alkaline/SDS Extraction of Crude Genomic DNA (gDNA), Common

1. NaOH stock solution, 1 M in water, stored in a plastic bottle.
2. SDS stock solution, 5% in water.
3. Alkaline/SDS solution: 20 mM NaOH, 0.1–0.2% SDS (*see* **Note 3**). Prepare immediately before use (*see* **Note 4**).
4. Tight snap cap 1.5 mL tube (*see* **Note 5**).
5. Dry thermo-block incubator.
6. Disposable rubber gloves.
7. Vortex mixer.

2.2 Sampling of Arabidopsis Leaf, Seed Pool

1. Small and sharp scissors, such as ophthalmic scissors.
2. Tweezers with sharp and smooth tips.
3. Kimwipe wiper paper.
4. Water for wipe wetting.
5. A digital camera or cell phone that takes pictures to remember individual plant placement.
6. Seed-spoon 50 (Bio Medical Science Inc.).

2.3 High-Throughput Culture and gDNA Extraction of Yeast

1. 96-well 2 mL deep-well plate, sterilized.
2. Gas-permeable plate seal for the deep-well plate culture.
3. Plate shaker and incubator for culture.
4. Multichannel pipette (8ch or 12ch).
5. PCR plate (for high-throughput extraction).
6. PCR plate seal.
7. Plate centrifuge.
8. Thermal cycler.

2.4 CD-PCR and Post-PCR Processing

1. 100 mM α-CD stock: This is 10× concentration for PCR. Weigh 3.89 g α-cyclodextrin, then prepare a 40 mL solution in a disposable 50 mL tube. Fill up the solution by using the scale printed on the tube. Dispense aliquots into screw-cap tubes. Store the working tube at 4 °C and the rest in the freezer (*see* **Note 6**).
2. 60%w/v sucrose stock: Weigh 24.0 g sucrose, then prepare a 40 mL solution in a disposable 50 mL tube. Dispense aliquots, store the working tube at 4 °C and the rest in the freezer.
3. Tracking dye stock solution: Acid Red 27, 2% in water (*see* **Note 7**).
4. DNA polymerase for PCR.
5. PCR buffer.
6. dNTP mixture.
7. Gene-specific primers.
8. 0.2 mL PCR tube.
9. 0.2 mL PCR plate.
10. PCR plate seal.
11. Aluminum PCR tube/plate rack for an ice bath.
12. Thermal cycler.
13. 5× T7-buffer with red loading dye: 40% sucrose, 100 mM Tris–HCl, pH 8.0, 500 mM NaCl, 5 mM $MgCl_2$, 0.10% Acid Red 27.
14. Multichannel micropipette (for high-throughput handling, 10 or 20 μL size, 12ch or 8ch).

2.5 Agarose Gel Electrophoresis

1. Agarose gel electrophoresis chamber, gel preparation equipment (*see* **Note 8**).
2. Electrophoresis buffer: 0.5× TBE buffer supplemented with 0.5 mg/L EtBr (*see* **Note 9**).
3. Agarose gel: suitable agarose concentration according to the size of the PCR fragment, 0.5× TBE buffer supplemented with 0.5 mg/L EtBr (*see* **Note 9**).

3 Methods

3.1 gDNA Extraction from Arabidopsis Leaf by Hot Alkaline/SDS Method

This protocol describes handling the typical model plant *Arabidopsis thaliana*, but it can also be applied to other plants [1].

1. Plan the size of the experiment for the day (number of samples, amount of reagents needed, time required).
2. Prepare the alkaline/SDS (20 mM, 0.1%) solution to be used on the day.

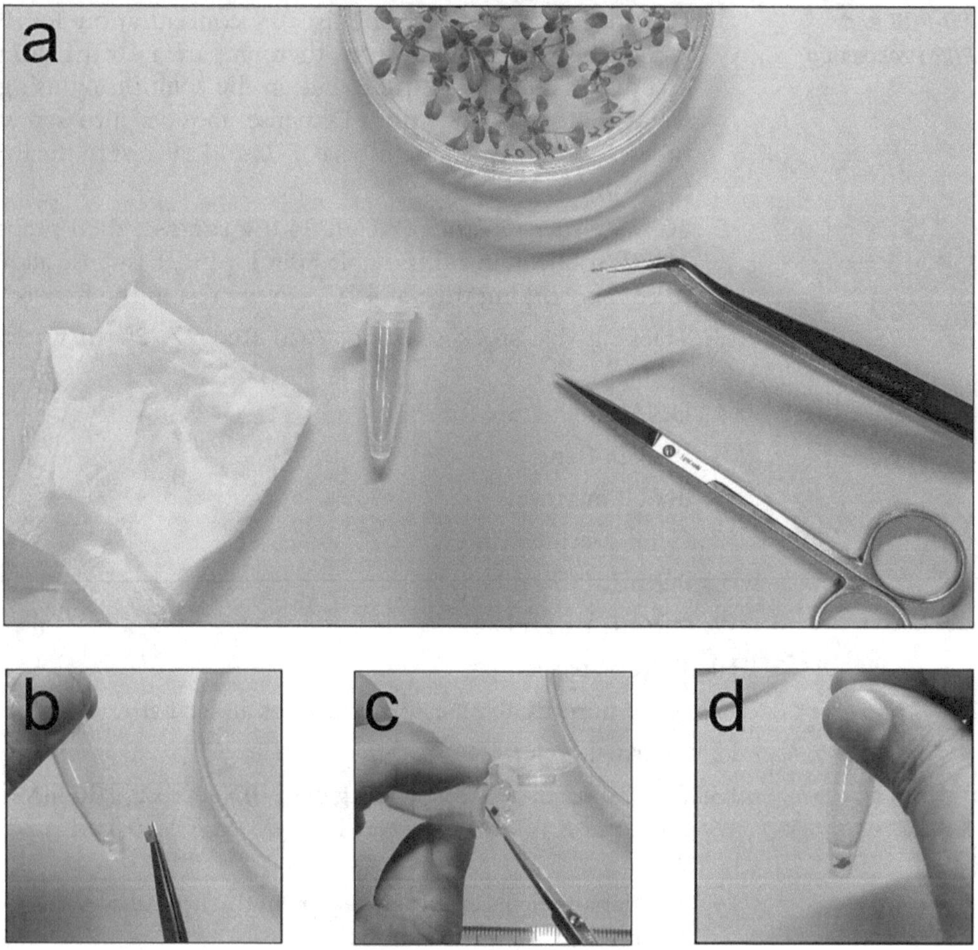

Fig. 1 Sampling of Arabidopsis leaves for genomic DNA extraction. (**a**) Set of items needed for leaf sampling. Two-week-old *Arabidopsis* individuals for genotyping are first labeled on the dish surface with an identification symbol/number. The required materials are wet Kimwipe, 1.5 mL tubes labeled and dispenced with the alkaline/SDS solution, scissors, and tweezers. (**b**) Cut 3–4 mm leaf tissue from the tip of rosette leaf of the *Arabidopsis* using sharp scissors. (**c**) Transfer the tissue into the tube while holding it with scissors. (**d**) Immerse the leaf sample in the alkaline/SDS solution quickly

3. Dispense 100 μL of the solution into tight snap cap 1.5 mL tubes (*see* **Note 5**).
4. Label individual plants with a simple number/symbol. For example, "101," "A02," etc. Label tubes with the same notation. Take a photo of sample plants so that there are no mix-ups later.
5. Cut 3–4 mm from the tip of an Arabidopsis rosette leaf by using ophthalmic scissors (Fig. 1) (*see* **Note 10**). Put the piece of leaf in a tube, drop it down, and immerse it in the solution. If it gets caught in the middle, flick the tube off with your finger (or put it into the solution with tweezers). Clean the tips of the scissors

with a Kimwipe paper dampened with water, and move on to the next sample (*see* **Note 11**).

6. Wear disposable rubber gloves and protective goggles for safety (*see* **Note 12**).
7. Put the tube on a dry thermo-block incubator, then heat the tube to 95 °C for 5 min.
8. Optionally, Vortex the hot tube at maximum speed for 3–5 s (*see* **Note 13**).
9. Move the tube to the crushed ice bath. The supernatant is now ready for CD-PCR (*see* **Note 14**).
10. Optionally, transfer the supernatant to an 8/12/96-well format tube/plate if multiple genotyping experiments are required (*see* **Note 15**).
11. For prolonged storage, the gDNA sample can be stored below −20 °C until required for use.

3.2 gDNA Extraction from Arabidopsis Seed Stock

1. Prepare the solution of the alkaline/SDS (20 mM, 0.2%) for the day.
2. Dispense 100 μL of the solution into tight snap cap 1.5 mL tubes (*see* **Note 5**).
3. Measure about 50 grains of *Arabidopsis* seed by seed-spoon 50.
4. Put it in a tube, drop it down, and immerse it in the solution.
5. Wear disposable rubber gloves and protective goggles for safety (*see* **Note 12**).
6. Put the tube on a dry thermo-block incubator, then heat the tube to 95 °C for 10 min.
7. Optionally, Vortex the hot tube at maximum speed for 3–5 s (*see* **Note 13**).
8. Transfer the tube to a crushed ice bath.
9. Spin down the tube. The supernatant is ready for CD-PCR (*see* **Note 14**)

3.3 High-Throughput Style Culture and gDNA Extraction from Yeast

In the case of microorganisms that can be cultured in liquid form, such as yeast, gDNA preparation, CD-PCR, and agarose gel electrophoresis can all be performed in a high-thorough style (Fig. 2).

1. Culture yeast clone using a 96-well format 2 mL deep-well plate covered by a gas-permeable seal.
2. Prepare the alkaline/SDS solution.
3. Dispense 40 μL of the solution into a PCR plate (vessel for gDNA extraction).
4. Transfer 2 μL of the culture to the plate containing the alkaline/SDS solution by using a multichannel pipette. Mix the

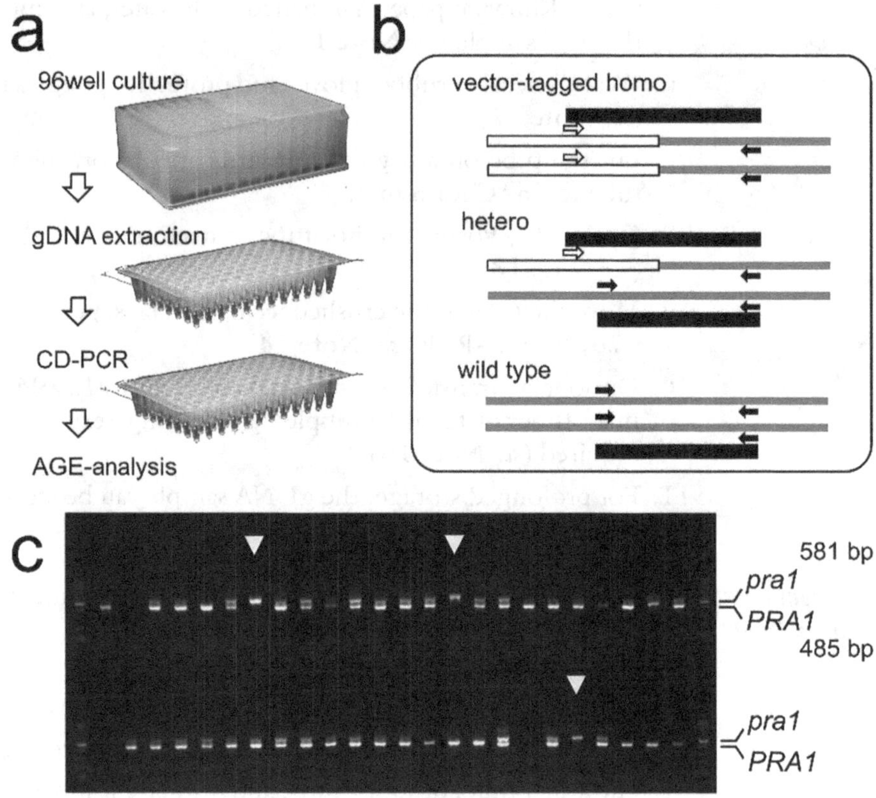

Fig. 2 High-throughput CD-PCR genotyping of yeast *Ogataea polymorpha*. The *PRA1* gene of *O. polymorpha*, a yeast that undergoes homothallic conjugation and polyploidy changes, was disrupted by homologous recombination using a vector-tagged gene-fragment. Since the yeast also has high NHEJ activity, the vector is often integrated into unintended loci. We tested whether the *PRA1* locus had been completely replaced by the deletion gene (*pra1*). (**a**) Workflow of genotyping. Candidate yeast clones were cultured in a 96-well deep-well plate. gDNA was extracted as 96-well style, then 0.3 μL of the crude gDNA was subjected to 18 μL of the three-primer CD-PCR (PRA1ko.F, TACAAACCTGTTCTGAATAGAGAGTCACC; PRA1ko.R, GGTGTCGAGGATAACCTT GAACGAC ; DELvec_EM7.R, CCTCACCTTGTCGTATTATACTATGCCG). PCR was performed with an initial denaturation step at 95 °C 2 min followed by 40 cycles of (97 °C 10 s, 55 °C 20 s, 72 °C 1 min). (**b**) Design of three primer genotyping PCR. Gray lines indicate wild-type genes, white lines indicate vector genes, and arrows indicate specific primers. The black bar means PCR products, and the length of the PCR for the vector-tagged deletion allele (581 bp) is designed to be 20% larger than that of the wild-type gene (485 bp). (**c**) Agarose gel electrophoresis image of the CD-PCR genotyping. Triangles indicate homo deletion lines

sample by stirring the multichannel pipette tip. The remaining culture should be refrigerated until the genotyping is performed or frozen for later recovery.

5. Cover the plate with a PCR plate seal and heat at 95 °C for 5 min in a thermal cycler, then cool down to 4 °C. Spin down the PCR plate by using a plate centrifuge (*see* **Note 14**).

Table 1
Examples of CD-PCR premix

Premix	ExTaq base	KOD FX Neo base	Home-made "B1"
Buffer	1× *ExTaq* buffer (TaKaRa)	1× PCR buffer for KOD FX Neo (Toyobo)	60 mM Tris–HCl (8.8) 20 mM $(NH_4)_2SO_4$ 2 mM $MgSO_4$
Substrate	0.2 mM dNTP	0.4 mM dNTP	0.2 mM dNTP
Cyclodextrins	10 mM α-CD	10 mM α-CD	10 mM α-CD 3 mM G_1-β-CD
Enzyme	0.025 U/μL *ExTaq* (TaKaRa)	0.02 U/μL KOD FX Neo (Toyobo)	*0.04 U/μL *Taq* *0.0025 U/μL *Pfu*
Dense solute dye	12% sucrose 0.02% acid red 27	– 0.01% cresol red-Na	10.9% sucrose 0.005% BPB-NaOH

Blend-type PCR enzymes (e.g., *ExTaq*, TaKaRa; One*Taq*, NEB) are routinely used for CD-PCR premix. KOD FX Neo-based PCR premix is powerful for crude gDNA that contain unknown inhibitors. "B1" premix is prepared using cheap labstock enzymes (recombinant *Taq* and *Pfu*) as described in [1].

* The concentrations are for the home-made enzymes (*Taq* plus *Pfu*), because they are cheap but relatively unreliable in terms of activity values. Taq 0.025 U/uL Taq plus 0.0015 U/uL Pfu is recommended when using commercial enzymes.

3.4 CD-PCR

The trick of CD-PCR is only to add 10 mM α-CD at the final concentration to a PCR premix of any composition (*see* **Note 16**). Preparing the PCR premix containing a weighting reagent (e.g., sucrose) and an electrophoresis migration dye is recommended because they support direct subjection of the CD-PCR products to an agarose gel electrophoresis (*see* **Note 17**). Table 1 shows examples of CD-PCR composition (without primers).

1. Plan the composition of the PCR premix and the thermal program. Follow the manufacturer's instructions for the reaction buffer, dNTP, DNA polymerase enzyme, and primers concentration. Add 10 mM α-CD, 8–12% sucrose, and tracking dye (e.g.., 0.02% Acid Red 27) at final concentrations to the recipe (*see* **Note 17**). The liquid volume is adjusted with ultrapure water. Design a thermal program according to the manufacturer's instructions (*see* **Note 18**).
2. Prepare the PCR premix without primers on ice, adding reagents in order and mixing step by step: ultrapure water, the reaction buffer, dNTPs, sucrose, α-CD, DNA polymerase, and the tracking dye. Optionally, the PCR remixes at this stage can be frozen and stored (*see* **Note 19**). If you are using a frozen PCR premix, melt it in water while monitoring the ice in the tube to prevent the premix from becoming warm, and leave it on the ice for 5 min after melting.

3. Add the target gene's primer pair (or trio) to prepare a working CD-PCR premix.
4. Pre-warm the thermal cycler and set the thermal program at this step (*see* **Note 20**).
5. Set the aluminum rack on the ice bath and place the PCR tubes/plates on it. Dispense 15–20 μL of the working premix into the PCR tubes/plates.
6. Using a micropipette (or a multichannel micropipette), suck in 0.3–0.6 μL of crude gDNA solution for the sample (*see* **Note 21**). Visually check for the presence of the gDNA solution. Then, add the solution to the working PCR premix. At this point, press the plunger of the micropipette to the first stop. Mix for 2 s by stirring the pipette tip. Next, push the plunger in a little from the first stop to drain out the remaining small amount of liquid. Do not push the plunger too far, as air bubbles will form. Replace with a new tip and move on to the following sample.
7. Seal the PCR tube/plate. Incubate for 3 min on ice. If bubbles are in the reaction solution, spin down the PCR tube/plate in a centrifuge.
8. Run the thermal program of the PCR.
9. Load 2–5 μL of the CD-PCR product directly in agarose gel supplemented with EtBr (*see* **Notes 9** and **22**).
10. Run agarose gel electrophoresis (AGE), then analyze the gel image.

3.5 Post-processing of CD-PCR Product by Restriction Enzyme

If there are differences in the restriction enzyme sites between the alleles being examined, the CD-PCR product can be treated with a restriction enzyme to detect RFLP [1].

1. Prepare 2× restriction enzyme premix: 2× enzyme buffer, 0.3 U/μL restriction enzyme, 8% sucrose, 0.03% Acid Red 27 (*see* **Note 23**).
2. Dispense 5 μL of the 2× restriction enzyme premix into 0.2 mL PCR tubes/plate.
3. Add 5 μL of the CD-PCR product. Mix by pipetting.
4. Incubate the tube at 37 °C for 2 h.
5. Load 3–10 μL of the PCR/restriction enzyme digestion product in agarose gel.
6. Run agarose gel electrophoresis, then analyze the gel image.

3.6 T7 Endonuclease I Assay of the CD-PCR Product

T7 endonuclease I assay can detect insertion/deletion mismatch alleles of 2 bp or more created by recent genome editing techniques (Fig. 3).

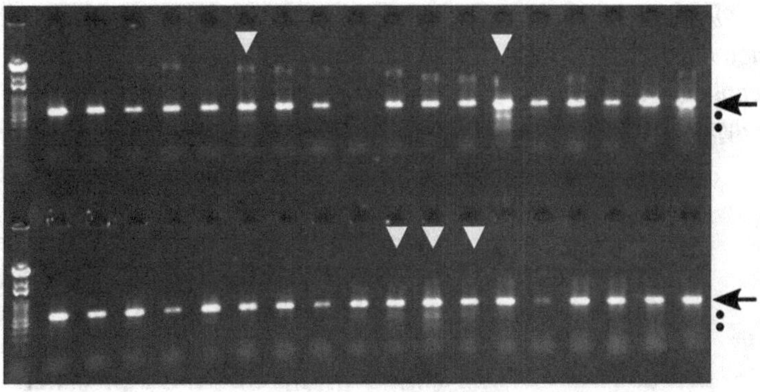

Fig. 3 Genotyping of small insertion/deletion allele of *Arabidopsis thaliana ABL2D* gene created by CRISPR/Cas editing. gDNA was extracted from the rosette leaves of 2-week-old plants using alkaline/SDS (20 mM/0.1%). 0.3 μL of the gDNA was mixed with 20 μL of CD-PCR mixture consisting of the homemade PCR mixture (B1) and gene-specific primer pairs (D.2_FW, CATGT CAGCATGTTACACGGC; D.2_RV, ACCCAATAATCCCTAAGAGCACG). PCR was performed with an initial denaturation step at 95 °C 5 min followed by 40 cycles of (98 °C 10 s, 51 °C 20 s, 72 °C 60 s), 72 °C 5 min. The thermal program of the PCR was directly linked to the program of the hybrid formation. Next, 5 μL of the PCR product was mixed with 5 μL of the 2× T7E1 enzyme premix. After incubation at 37 °C for 40 min, 5 μL of the T7E1 assay product was analyzed by 1.5% AGE. When there is a mismatch hybridization product of the allele PCR product, two bands of the T7 endonuclease I degradation product (dots) appear below the primary band corresponding to the PCR (arrow), so we used this to screen the *ABL2D* genome-edited strain (triangle)

1. After regular CD-PCR, run the thermal program of hybrid formation: denature step 95 °C for 5 min, hybrid annealing lowering step to lower 95 °C to 85 °C (−2.0 °C/s) followed by 85 °C to 25 °C (−0.1 °C/s). Optionally, reference PCR product DNA could be added to the hybrid formation reaction for the detection of mutant allele homozygous genotypes (*see* **Note 24**).
2. Prepare 2× T7E1 enzyme premix: 5× T7-buffer with red loading dye, 2× rT7E1 enzyme.
3. Dispense 5 μL of the enzyme premix into 0.2 mL PCR tubes/plate.
4. Add 5 μL of the CD-PCR product. Mix by pipetting.
5. Incubate the tube at 37 °C for 30 min.
6. Load 3–10 μL of the CD-PCR/T7E1 product directly in agarose gel.
7. Run agarose gel electrophoresis, then analyze the gel image.

3.7 Guideline for Design High-Throughput CD-PCR Genotyping

1. In the model experiments of CD-PCR, DNA of a length exceeding 2 kbp has been successfully amplified [1], but due to the complexity of gDNA and stability of the crude gDNA, the region to be amplified in routine genotyping should be 1 kbp or less, preferably 500 bp or less.
2. It is not advisable to make a genotyping judgment based on the absence of a PCR band. It is difficult to eliminate experimental problems in high-throughput work completely, and it is not possible to distinguish whether a no PCR band means that the allele type does not amplify with the primer pair or whether it simply means that the PCR failed. Design a set of primers that amplify all opposing alleles specifically and with similar amplification efficiency in the same thermal cycle. PCR using three (or more) primers can avoid the problem of no bands appearing (Fig. 2).
3. To facilitate differentiation of allele-specific three-primer PCR bands on agarose gel electrophoresis, the amplified size of the PCR should differ by approximately 20–25% (e.g., 400 bp vs. 500 bp, 200 bp vs. 250 bp). It is preferable to make the amplified region of "the allele you want to detect" longer.
4. If it is challenging to create differences in multi-primer CD-PCR size between alli, please consider combining it with experimental techniques that detect restriction enzyme site polymorphisms and insertion/deletion polymorphisms as a post-process of CD-PCR (Fig. 3).

4 Notes

1. In ordinary laboratories, reagent stocks, equipment such as dispensing spoons, beakers, measuring cylinders, and purified water stored in bottles are occasionally contaminated with the gene you wish to analyze. Note that even the slightest contamination can cause false positive PCRs.
2. The reagents should be purchased separately from the routine reagents and kept as "PCR-specific" so that they will not be used for any other purpose. The reagent grade can be an analytical grade.
3. An increase in SDS concentration is expected to improve DNA extraction efficiency. On the other hand, the concentration of endogenous inhibitors contained in the crude extract also increases, so we recommend investigating the concentration in a preliminary test for the sample.
4. Low concentrations of alkali are easily neutralized by carbon dioxide in the air.

5. We recommend the 1.5 mL tube with a tight snap cap (not soft-touch, not easy-open) to prevent the tube from bursting in the heat extraction step. INA OPTICA ST-0150F is an example. The usual microtube cap rocker is more labor-intensive but less effective in preventing bursts.
6. Precipitation may occur in 100 mM α-CD stock solution in both refrigerated and frozen storage. Before use, dissolve it by heating it at 95 °C and then cooling it on ice. The solution should be stored in screw-cap tubes that can endure increased internal pressure during the heating.
7. We have recently preferred Acid Red 27 (0.02%) for the tracking dye of the PCR premix because it shows the migration front of agarose gel electrophoresis. Other popular tracking dyes, 0.005% bromophenol blue-NaOH, 0.005% cresol red-Na, and 0.01% orange G, are also workable with CD-PCR.
8. If you are planning high-throughput experiments, using an agarose gel electrophoresis chamber and gel preparation equipment that matches the pitch of the 8ch/12ch multichannel pipette is preferable.
9. We prefer TBE buffer because it has a higher resolution than TAE buffer when separating regular genotyping PCR fragments <2 kb by electrophoresis. We also like adding EtBr to the buffer and the agarose gel. The in-gel EtBr staining method is time- and labor-efficient. Staining the gel in a later process is also possible and increases the resolution of electrophoresis, but it is time-consuming.
10. Keep the size of the plant tissue to be treated with alkaline/SDS. If the sample size is too large, the experiment's reproducibility will decrease. When plants grown aseptically are to be used for genotyping, sampling of plant leaves is carried out inside a clean bench.
11. It is not necessary to change the wiper paper for every sample. Slight cross-contamination in this step has little effect on experimental results. Replace when you can see the green stain.
12. Hot alkaline/SDS solution is harmful to the skin and eyes. Using a 1.5 mL tube with a tight snap cap prevents most, but not all, accidents due to bursts in the heat step.
13. Vortexing the hot tube for 3–5 s and transferring it onto ice improves PCR reproducibility.
14. The gDNA samples stored on ice or in cold storage at 4 °C are good as templates for CD-PCR during the day, but the reproducibility of the CD-PCR is poor after the next day. Multiple freeze/thaw cycles also deteriorate sample quality. Degradation of template quality can be restored by re-heating and rapidly cooling the crude gDNA sample.

15. 8/12/96-well style gDNA extraction is NOT recommended for plant samples due to difficulty in handling.
16. The protective effect of SDS is sufficient with α-CD alone. In some cases, the effect of inhibitory substances in the sample can be alleviated by adding 1-3 mM of water-soluble β-CD derivatives or γ-CD in addition to α-CD. However, some CD species have an inhibitory effect on PCR, so this should be investigated in a pilot study [1].
17. If you use a ready-made PCR premix containing a weighting reagent and dye, adding these reagents is unnecessary.
18. The Tm of the primer oligo DNA may change (increase) due to the influence of the CD and the weighting reagent. A small alkaline change in reaction slightly affects the Tm. It is advisable to determine the workable annealing temperature of PCR in a pilot study.
19. PCR premix without primer can withstand about ten freeze-thaw cycles.
20. If the PCR premix containing gDNA and primers is left for a long time before starting the PCR, the reproducibility of the PCR may decrease.
21. Due to the viscosity of the gDNA solution and the effect of the multichannel micropipette, the accuracy of the liquid volume operation will decrease so that the amount will be set at about 1/50 to 1/30 of the working PCR premix. Excessive amounts of gDNA extract reduce the reproducibility of experiments.
22. If you use a multichannel pipette for loading PCR product to the agarose gel, check the straightness and pitch of pipette tips before sucking in the sample. If the alignment of the tips is bent or the pitch is unnatural, adjust them. After loading the sample, wash the tips by pumping the electrophoresis buffer, absorb any excess droplets on a paper towel, and reuse them for the following sample. Replacing the tip for each sample reduces work efficiency.
23. The buffer for the restriction enzyme is designed considering the salt concentration in the PCR reaction solution that is brought into the reaction.
24. Check the amount of target-PCR product first, then add reference PCR product according to the amount. Reference (wild-type homo- or mutant-In/Del-homo-) PCR products should be prepared with the same composition as CD-PCR.

Conflict of Interest

The authors have no conflicts of interest to declare that are relevant to the content of this chapter.

References

1. Nakanishi Y, Kawashima T, Naganawa M, Mikami T, Maeshima M, Ishiguro S (2022) Letter to the editor: Speedy plant genotyping by SDS-tolerant cyclodextrin-PCR. Plant Cell Physiol 63(8):1025–1028
2. Dos Santos Silva Araújo L, Lazzara G, Chiappisi L (2021) Cyclodextrin/surfactant inclusion complexes: an integrated view of their thermodynamic and structural properties. Adv Colloid Interf Sci 289:102375
3. Brocos P, Banquy X, Díaz-Vergara N, Perez-Casas S, Piñeiro A, Costas M (2011) A critical approach to the thermodynamic characterization of inclusion complexes: multiple-temperature isothermal titration calorimetric studies of native cyclodextrins with sodium dodecyl sulfate. J Phys Chem B 115:14381–14396
4. Vouillot L, Thelie A, Pollet N (2015) Comparison of T7EI and surveyor mismatch cleavage assays to detect mutations triggered by engineered nucleases. G3 (Bethesda) 5:407–415

Chapter 9

Probe-Based qPCR Workflow Using Fast Nucleic Acid Sampling and Molecular Genotyping in a Dioecious Plant

Anthony Torres and Reginald Gaudino

Abstract

This chapter presents a protocol for probe-based quantitative polymerase chain reaction (qPCR) genotyping, focusing on male-specific genotype identification in plant samples. DNA is isolated from raw tissue or FTA cards and subjected to a preamplification enrichment reaction. The diluted product is then used as input for qPCR detection, employing specific probes to identify the presence or absence of the target genotype. This robust and high-throughput method offers increased specificity for genotyping various sample types, including crude nucleic acid extracts. An advantage of qPCR is that genotyping can be performed on various sample types, including from crudely prepared nucleic acid extracts used as input with inhibitor-tolerant polymerases. In this workflow, we describe an extraction and probe-based qPCR method used to detect male-specific genotype identification in plant sample DNA extracted from raw tissue or FTA imprinted cards.

Key words DNA, Amplicon, PCR, Preamplification, qPCR , Primers, Probe, Cycle threshold, Detection, qPCR, Genotyping, Probe-based detection, Male-specific genotype, Plant samples, FTA cards, Preamplification enrichment

1 Introduction

Real-time quantitative PCR (qPCR) is a reliable molecular method for genotyping individual and pooled samples of various tissue types and input preparation [1]. Of the molecular detection methods, with good design and execution, it is robust, precise, and accurate. Specific detection of target loci using a molecular probe can be used to assess genotypes within a population [2]. Coupling polymerase chain reaction with a detection probe allows for real-time amplification analysis over cycles, allowing the end user to be accurately informed about target specificity, selectivity, and sensitivity [3]. The selectivity of the probe is flexible and user-designed for any target range of genotyping applications. Ensuring additional specificity in design is essential [4]. Some probes may be designed for presence/absence variation genotyping, where the

Anthony Torres (ed.), *Genotyping: Methods and Protocols*, Methods in Molecular Biology, vol. 2943,
https://doi.org/10.1007/978-1-0716-4642-7_9,

presence or absence of that genotype is equally informative [5–7]. In our analysis, we developed a male-specific qPCR primer and probe set that binds to a Y chromosome male-specific amplicon region from *C. sativa* samples [8]. *C. sativa* is dioecious, and female inflorescence tissue is more productive without pollen [9, 10]. Ensuring a male-free environment is a critical agricultural application in this context [11]. Detecting specific target probes observed only in a specific genotype, such as male unique regions in the Y chromosome, allows for highly accurate molecular diagnostic analysis [8, 12]. Here, we present a high-throughput protocol for DNA isolation from tissue or DNA-imprinted FTA card material, which is used as input into an enriched quantitative PCR workflow.

2 Materials

2.1 Sample Preparation

1. Bunsen burner.
2. Calibrated p1, p100, p200 multichannel pipettes and p1, p200, p1000 single-channel pipettes.
3. 100–1000 μL nuclease-free aerosol filtered pipette tips.
4. 20–300 μL nuclease-free aerosol filtered pipette tips.
5. 1–100 μL nuclease-free aerosol filtered pipette tips.
6. 0.1–20 μL nuclease-free aerosol filtered pipette tips.
7. Stainless steel dissecting forceps.
8. 250 mL beaker.
9. Optional: plate piercer.
10. 1.5 mL centrifuge tubes.
11. Gloves.
12. Isopropanol.
13. Nuclease-free water.
14. Extraction buffer solution: 0.25 M Tris–HCl, pH 8, 0.25% Triton-X.
15. FTA plant saver cards.
16. Unicore punch kit, 2 mm.

2.2 Preamplification Reaction

1. Standard PCR thermocycler.
2. Mini plate spinner.
3. Gloves.
4. Clean designated space.
5. Clear 96-well plate or similar.
6. Sterile loading trough.

7. Genotype-specific qPCR oligonucleotides (<300 bp targets) (*see* **Note 1**).
8. GoTaq PCR Master Mix (Promega).
9. Genotype-specific positive DNA for the positive control template.
10. AlumaSeal II sealing foils.

2.3 qPCR Reaction

1. Standard real-time PCR thermocycler.
2. Genotype-specific qPCR oligonucleotides (<300 bp targets) (*see* **Note 1**).
3. Genotype-specific 5′ fluorophore with 3′ quencher qPCR oligonucleotides (*see* **Note 2**).
4. QuantiNova Probe qPCR Kit (Qiagen).
5. Genotype-specific positive DNA for the positive control template.
6. Optical sealing foils.

3 Methods

3.1 Sample Preparation

1. Prepare the workstation on the lab bench by wiping down the surface with alcohol where the preparation will be performed.
2. Prepare clean PCR setup with a clean 96 well plate for sample preparation in an easily accessible area.
3. Sterilize forceps by dipping them in alcohol and flame sterilizing them or aseptically wiping excess alcohol until all alcohol has evaporated.
4. Remove a small sampling of raw tissue pieces or imprinted FTA card (~2 mm) with flame-sterilized forceps or alcohol-sterilized QIAGEN 2 mm UniCore punch tool, taking care not to touch the sample directly with gloves.
5. Carefully place the sample into the PCR plate at the bottom of the well so that no contact is made with the top of the PCR plate.
6. Repeat **steps 3–5** with remaining lots/replicates until all samples are processed.
7. Fill sterile liquid basin/trough with Extraction Buffer using dedicated single-channel or multichannel 30–300 μL multichannel pipettor loaded with clean tips to draw and dispense 100 μL of extraction buffer into each well.
8. Repeat for all remaining samples and replace tips each time. After, seal with AlumaSeal II film strip.

9. Spin plates down in plate spinner for 20 s to ensure all material is fully immersed in the extraction buffer.
10. Load the plate on the thermocycler and set it to run for 5 min. Boil with the lid closed and the heated cover on.
    (a) ~95–100 °C 5 min 1×.
11. Prepare a clean 96-well plate(s) for performing a serial dilution with the boiled sample.
12. Following boil, fill ~10 mL of nuclease-free water into a new sterile liquid basin/trough and pipette the appropriate dilution volume into a clean PCR plate for a 1:100 dilution (e.g., 90 μL Water +10 μL boiled sample serial diluted with 1:10 10 μL boiled diluted sample + 90 μL Water for 1:100 dilution).
13. Spin plates down in plate spinner for 20 s to ensure all material is fully immersed in the extraction buffer.
14. Using a plate piercer or pipette tips, pierce the boiled extraction plate.
15. Perform a 1:100 dilution in dilution PCR plate(s) with the boiled sample.

3.2 Preamplification Enrichment Reaction Setup

1. Prepare a clean PCR setup with a clean 96-well plate for sample preparation in an easily accessible area.
2. In a dedicated pre-PCR space, prepare a PCR master mix according to the following:
   Reagent:
   GoTaq PCR master mix: 12.5 μL.
   Genotype-specific primer mix (10 μM): 1 μL.
   Nuclease-free water: 7 μL.
3. Prepare pre-Amp master mix in a sterile multichannel loading trough.
4. Set up a clean, clear 96-well plate.
5. In the dedicated PCR hood, use a 0.1–10 μL multichannel pipette, load four to eight 0.1–20 μL filtered pipette tips, pipette 20 μL of the pre-Amp master mix to each column of the plate, respectively, for columns 1–11.
6. Eject tips into waste container.
7. For all reactions, using a 0.1–10 μL multichannel or single channel pipette, draw and dispense 5 μL of 1:100 diluted crude extract from crude extraction plate to corresponding wells of the preamplification plate and eject tips after dispensing (*see* **Note 3**).
8. Using an applicator, seal the plate with AlumaSeal II plate sealing film foil (sticky side down) and ensure that all wells are sealed.

9. Spin down in a mini plate spinner for 10 s.
10. Place the pre-amp plate in the thermocycler and select and run the Pre-Amp Enrichment PCR protocol:
    (a) 95 °C for 5 min 1×.
    (b) 95 °C for 15 s.
    (c) 60 °C for 30 s (*see* **Note 5**).
    (d) 72 °C for 30 s.
    (e) Repeat **steps a–d** 10×.
    (f) 4 °C hold.
11. Following the preamplification reaction, fill ~10 mL of nuclease-free water into a new sterile liquid basin/trough.
12. Spin preamplification plates down in plate spinner for 20 s.
13. Using a plate piercer or pipette tips, pierce the wells of the preamplification enrichment plate.
14. Pipette 100 μL water to dilute the preamplification enrichment reaction to a 1:5 working dilution.

3.3 qPCR Reaction Setup

1. Prepare clean PCR setup with a clean 96-well plate for sample preparation in an easily accessible area.
2. Prepare a qPCR master mix according to the following:
   Reagent:
   QuantNova probe qPCR Master Mix 2×: 3.75 μL (*see* **Note 4**).
   Genotype-specific primer mix (10 μM): 1 μL.
   Genotype-specific probe mix (250 nM): 0.2 μL.
   Nuclease-free water: 0.2 μL.
3. In the dedicated PCR space, use a 0.1–10 μL multichannel pipette, load 0.1–20 μL filtered pipette tips, pipette 5 μL of the qPCR reaction mix to each well of the plate, respectively, for each sample.
4. Change tips and prepare to load samples, use fresh tips for each new sample.
   (a) For all PreAmp enrichment reactions, using a 0.1–10 μL multichannel pipette or single channel pipette, draw and dispense 5 μL of 1:5 diluted preamp solution to corresponding wells of qPCR plate, be sure tips are below surface of mix before dispensing, mix by pipetting up and down several time, then, eject tips after final mix.
   (b) For all reactions with purified genomic DNA, using a 0.1–10 μL multichannel pipette or single-channel pipette, draw and dispense 5 μL of kit (1–10 ng/μL)* purified genomic DNA from DNEasy or similar ensure tips are

below the surface of mix before dispensing, mix by pipetting up and down several times, then, eject tips after final mix.

5. Pipette to mix 1–3 times and eject tip.
6. To minimize bubbles in qPCR reaction plate, pipette pre-amped solution slowly (*see* **Note 6**).
7. After each sample well has been filled, obtain the control standards and aliquot 5 μL of positive control template, negative control template, and no template control into their each own appropriate wells.
8. Seal plate thoroughly with qPCR optical sealing foil.
9. Spin the plate for 10 s.
10. The plate is now ready for data acquisition.
11. Place the qPCR plate in the real-time instrument and select and run the QPCR/multiplex protocol:
    (a) 95 °C for 3 min 1×.
    (b) 95 °C for 15 s.
    (c) 60 °C for 60 s (With Plate Read/Data Acquisition) (*see* **Note 5**).
    Repeat **steps b** and **c** 40×.
12. Run qPCR reaction on qPCR instrument with appropriate data acquisition settings.

3.4 qPCR Instrument Setup

3.4.1 Instrument Run: Qiagen ROTOR-GENE Q

1. Double click application icon to open.
2. To create protocol: Select New Protocol → Select Quantinova PCR → Enter Tube and Reaction Size → Enter PCR parameters from **step 3.3.11**, ensuring plate read step at appropriate annealing step for assay (Table 1). Save protocol file → Click Run.
3. Enter all sample names as unknowns on the following screen and mark the PTC and NTC samples as such.

Table 1
qPCR/multiplex amplification protocol and thermal cycling steps

Step	Temperature (°C)	Time	Repetitions	Notes
a	95	3 min	1×	
b	95	15 s	40×	
c	60	60 s		Plate read/data acquisition (*see* **Note 5**)

The table details the temperature and duration of each step in the specified process, including a repeated cycle with data acquisition

4. After the run is complete, click the analyze button and select the color channel that you intend to analyze.
5. Save the reaction as the qPCR experiment barcode.
6. Example File Name Template: yyyymmdd_[Location] QPCRID###_initials_qPCR NOTE: Set a threshold for calling a Ct value and export the data to excel from the Rotor-gene software.
7. Once the run is complete, open Qiagen analysis software from the desktop.
8. Open datafile.
9. Select Analysis tab, select target color cycling from directory for analysis (Green = FAM) (Orange = ROX), etc.
10. To all wells and apply the following analysis functions:
11. Dynamic tube.
12. Slope correction.
13. Use auto threshold when able, however, if no standard curve is ran, set threshold between 0.03 and 0.05. Fluorescent target threshold can be adjusted manually at your supervisor's discretion during analysis, however in general the threshold line should be set to start to mid exponential phase of amplification curve.
14. Once threshold is set, observe Ct values measured for each sample in the results summary window for each fluorescent color.
15. Confirm the presence or absence of signal in the controls and then in the sample wells both by looking at the shape for the second derivative curves for each well and the software demarcation.
16. NOTE: Samples with a positive signal will have a numerical value after the Cp is calculated; negative signals will be blank.
17. Repeat for all targets.

3.4.2 Instrument Run and Analysis: Roche LightCycler

1. Double Click Application Icon to Open.
2. Open Light Cycler 480 software from the desktop.
3. Enter username: Admin.
4. Password: Roche480.
5. Open qPCR experiment file.
6. In Experiment Tab, below Overview, select Apply Template and follow directory: From System Admin → Templates → Run Templates. Select QPCR Test Protocol and click check mark.

7. In Sample Editor Tab, below plate grid, select apply template and follow directory: From System admin → Templates → Sample Templates. Select QPCR Test Format and click check mark.
8. In Sample Editor Tab, in the lower right corner, select Import to upload the instrument file from LIMS. From the drop-down menus, select "Pos" for the left column and "Sample Name" for the right column.
9. Ensure that Rel Quant Protocol is selected in step one, and both FAM (465–510 nm) and HEX (533–580 nm) excitation filters are selected.
10. Save the reaction as the qPCR Experiment barcode:

    Example File Name Template: yyyymmdd_[Location] QPCRID###_initials_qPCR
11. Select Start.
12. NOTE: If a window comes up requesting "Reasons for a Change," enter your initials and select the check mark.
13. Once the run is complete, click the Analysis tab and select the Absolute Quantification/Second Derivative Max analysis mode.
14. At the bottom of the page, click on the Filter tab and change the filter to HEX.
15. Select Calculate to display the TC/Reference Cps for the sample wells and for the controls (Fig. 1).
16. Confirm the presence or absence of signal in the controls and then in the sample wells both by looking at the shape for the second-derivative curves for each well and the software demarcation.
17. Wells with detected signal will be marked red and those with non-detectable signals will be green. If the software is unable to discern whether the signal was detected, the well will be colored blue.
18. During analysis, samples observed with a positive signal will have a numerical value after the Cp is calculated; negative signals will be blank.
19. In cases of unexpected non-detects for the Reference Cp: Confirm the assessment by looking at the second-derivative curve. If the curve is consistent with that of a non-detect, record the sample name in your lab notebook as the sample will need to be re-prepped.
20. Refer to notes for the re-prep procedure.
21. In cases where the software is unable to make a call: Select the Drop-down menu and change the analysis type from high confidence to high sensitivity. If the well color of the uncertain wells becomes either red or green, record the value in lab notebook.

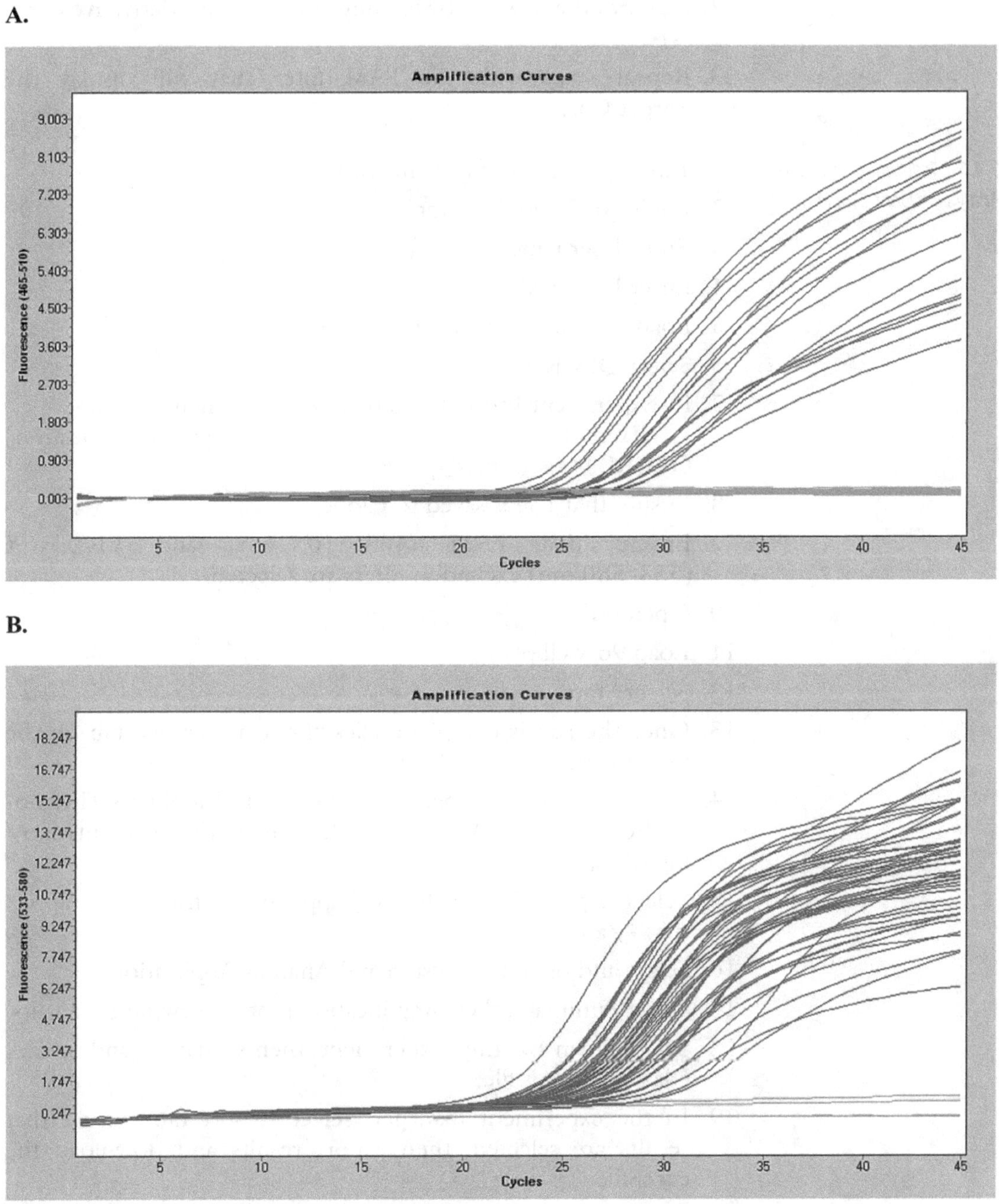

Fig. 1 Fluorescence analysis as measured by cycle on fluorescent channels for each probe (**a**) FAM and (**b**) HEX. FAM (465–510 nm) amplification curves correspond to Y chromosome sex determination and HEX (533–580 nm) amplification curves correspond internal positive control probe

22. Confirm the new call by looking at the second-derivative curve (s).
23. Repeat steps with the FAM filter that will display the Target Cps.

3.4.3 Instrument Run and Analysis: Quant Studio 5

1. Tap touchscreen of instrument to wake.
2. Login to Cloud Account.
3. Enter Username:
4. Enter Password:
5. Load experiment template from Table 1.
6. Select QPCR Test Method.
7. In Experiment Properties tab, enter experiment file as: Example yyyymmdd_[location] QPCRID###_initials_qPCR
8. Ensure that File is saved to Cloud.
9. Ensure that FAM (465–510 nm) and VIC/HEX (533–580 nm) excitation filters are selected.
10. Open tray.
11. Load 96-well plate.
12. Select Start.
13. Once the run is complete, click the Transfer File tab on the instrument to save archive data.
14. On the computer, open and browser and login to Thermo-Fisher Cloud Account: thermofisher.com/mysso/loginDisplay.
15. Select App Display: https://apps.thermofisher.com/apps/spa/#/apps.
16. Select and open the Design and Analysis Application.
17. Select Summary, then amplification plot to view target results.
18. Select Open Existing Experiment, then search for and open: .eds experiment file.
19. In the experiment manager, select Export tab, ensure that results are selected, then export results and download the excel file.
20. The file is downloaded as a zipped folder, unzip the folder and locate the QPCR test results excel file.
21. View the amplification curves in the Results Tab in the Design and Analysis application looking at the shape for the cycle amplification curves for each well and the software demarcation.

22. Each fluorescent target threshold can be adjusted manually at your supervisor's discretion during analysis, however in general the threshold line should be set to start to mid exponential phase of amplification curve.
23. Wells with detected signal above threshold line will be marked with Cycle quotient (Cq) value for Target 1 (FAM) and Target 2 (VIC/HEX), and those with non-detectable signals will be undetermined. Samples with a positive signal will have a numerical value for Target 1 (FAM) once the Cq is calculated, whereas negative signal will be undetermined.
24. In cases of unexpected non-detects or undetermined for the Target 2 (VIC/Hex): Confirm the assessment by looking at the amplification curve in the Results tab of the software.
25. If the curve is consistent with that of a non-detect, record the sample name in your lab notebook as the sample will need to be re-prepped.
26. Repeat steps with the Target 1 (FAM) filter that will display the Target 1 (FAM) Cq values.

3.4.4 Instrument Run: Biorad CFX96

1. Open Biorad CFX96 Maestro software from the desktop.
2. Enter Username:
3. Enter Password:
4. Double Click instrument icon to open.
5. To Create Protocol: Select File → New → Protocol → Enter PCR parameters from Table XX, ensuring plate read step at X annealing step. Save protocol file.
6. To create Plate: Select File → New → Plate → Select all wells and apply FAM and HEX Channel to all wells. (If Multiplex is used, Apply Cy5, ROX, and TAMRA channels additionally). → Apply Sample Type to appropriate wells → Save Plate file with unique qPCR experiment barcode.
7. To setup Run: Select Run → Under Protocol Load appropriate file. → Under Plate Load appropriate file. → Under Start Run → Select Open Lid and Load plate on to instrument → select Start Run.
8. Once the run is complete, save and archive datafile if not saved already.
9. In the Biorad CFX Maestro, open experiment datafile.
10. Confirm the presence or absence of signal in the controls and then in the sample wells both by analyzing the Target 2 (HEX) Cp data.
11. View the amplification curves in the Results Data Summary section looking at the shape/cycle of amplification curves for each well.

12. Each fluorescent target threshold can be adjusted manually at your supervisor's discretion during analysis, however in general the threshold line should be set to start to mid exponential phase of amplification curve.
13. Wells with detected signal above threshold line will be marked with Cp value for Target 1 (FAM) and Target 2 (HEX) and those with non-detectable signals will be undetermined.
14. NOTE: Samples with a positive signal will have a numerical value for Target 1 (FAM) once the Cp is calculated, whereas negative signal will be N/A.
15. In cases of unexpected non-detects or undetermined for the Target 2 (Hex):
16. Confirm the assessment by looking at the amplification curve in the Results tab of the software.
17. If the curve is consistent with that of a non-detect, record the sample name in your lab notebook as the sample will need to be re-prepped.
18. Repeat **steps 3.4.4.10–4.4.4.16** with the Target 1 (FAM) filter that will display the Target 1 (FAM) Cps.
19. Save excel file with a unique filename and archive.

4 Notes

1. qPCR oligonucleotides must be well designed to produce a single amplicon product generally not greater in length than 300 bp and compatible with thermodynamic properties of hybridization with a single-stranded fluorescently labeled oligonucleotide probe. Oligonucleotide primer should target conserved regions surrounding a target genotype of interest or be uniquely designed for a given genotype target.
2. Ensure that the probe oligonucleotides are fluorescently labeled to contain a 5′ end with a unique spectral range fluorophore and a 3′ quencher. Designs for probes should be genotype target-specific to yield a presence/absence result for a specified genotype, do not possess cross-reactive overlap of an unintended target, and should be checked for homology.
3. When pipetting from the extraction plate of crude lysis of fast nucleic acid extraction, ensure to pull from clear supernatant pipetting below the meniscus.
4. The annealing temperature for the designed oligonucleotide prepared for the target PCR primer mix should be at least 60 °C.

5. This protocol recommends a specific master mix. However, other vendors' inhibitor-tolerant qPCR master mixes can be utilized.
6. Bubbles in PCR plates can induce qPCR artifacts in data analysis and should be avoided/monitored.

Conflict of Interest

The authors have no conflicts of interest to declare that are relevant to the content of this chapter.

References

1. Sidstedt M, Radstrom P, Hedman J (2020) PCR inhibition in qPCR, dPCR and MPS-mechanisms and solutions. Anal Bioanal Chem 412(9):2009–2023. https://doi.org/10.1007/s00216-020-02490-2
2. Tran HT, Schramm C, Huynh MM, Shavrukov Y, Stangoulis JCR, Jenkins CLD, Anderson PA (2023) An accurate, reliable, and universal qPCR method to identify homozygous single insert T-DNA with the example of transgenic rice. Front Plant Sci 14:1221790. https://doi.org/10.3389/fpls.2023.1221790
3. Adams G (2020) A beginner's guide to RT-PCR, qPCR and RT-qPCR. Biochemist 42(3):48–53. https://doi.org/10.1042/BIO20200034
4. Hays A, Islam R, Matys K, Williams D (2022) Correction: Best practices in qPCR and dPCR validation in regulated bioanalytical laboratories. AAPS J 24(3):55. https://doi.org/10.1208/s12248-022-00705-1
5. Liu J, Zhang X, Dou L, Li W, Zhou X, Liu Y, Pei X, Ren Z, Zhang W, Li H, Wang W, Zou C, He K, Zhang F, Ma W, Shang H, Li H, Yang D, Xiao G, Ma X (2020) Patterns of presence-absence variants in Upland cotton. Sci China Life Sci 63(10):1600–1603. https://doi.org/10.1007/s11427-019-1602-y
6. Lee JS, Jahani M, Huang K, Mandel JR, Marek LF, Burke JM, Langlade NB, Owens GL, Rieseberg LH (2022) Expression complementation of gene presence/absence polymorphisms in hybrids contributes importantly to heterosis in sunflower. J Adv Res 42: 83–98. https://doi.org/10.1016/j.jare.2022.04.008
7. Clare SJ, King RM, Tawril AL, Havill JS, Muehlbauer GJ, Carey SB, Harkess A, Bassil N, Altendorf KR (2023) An affordable and convenient diagnostic marker to identify male and female hop plants. G3 (Bethesda) 14(1). https://doi.org/10.1093/g3journal/jkad216
8. Torres A, Pauli C, Givens R, Argyris J, Allen K, Monfort A, Gaudino RJ (2022) High-throughput methods to identify male Cannabis sativa using various genotyping methods. J Cannabis Res 4(1):57. https://doi.org/10.1186/s42238-022-00164-7
9. Flajsman M, Slapnik M, Murovec J (2021) Production of feminized seeds of high CBD Cannabis sativa L. by manipulation of sex expression and its application to breeding. Front Plant Sci 12:718092. https://doi.org/10.3389/fpls.2021.718092
10. Todd J, Song H, Van Acker R (2022) Does pollination alter the cannabinoid composition and yield of extracts from hemp (Cannabis sativa L. cv. Finola) flowers? Ind Crop Prod 183:114989. https://doi.org/10.1016/j.indcrop.2022.114989
11. Suchoff DH, Inoa SH, Stack GM, Wares AJ, Snyder SI, Murdock MJ, Rose JKC, Smart LB, Caton TA, Pearce RC (2024) Characterization of agronomic performance and sterility in triploid and diploid cannabinoid hemp. Agron J 116:2470. https://doi.org/10.1002/agj2.21618
12. Toth JA, Stack GM, Cala AR, Carlson CH, Wilk RL, Crawford JL, Viands DR, Philippe G, Smart CD, Rose JKC, Smart LB (2020) Development and validation of genetic markers for sex and cannabinoid chemotype in Cannabis sativa L. GCB Bioenergy 12(3): 213–222. https://doi.org/10.1111/gcbb.12667

Chapter 10

Probe-Based RT-qPCR Workflow for Dual Interrogation of Virus/Viroid Genotype in Molecular Diagnostics Multiplex

Anthony Torres and Reginald Gaudino

Abstract

This chapter presents a protocol for dual interrogation reverse transcriptase quantitative polymerase chain reaction (RT-qPCR) genotyping, a rapid, high-throughput, and economical molecular diagnostic approach for the simultaneous detection of viroid and virus genotypes in plant samples. The method utilizes crude nucleic acid extracts from raw tissue or FTA cards, streamlining sample preparation and enabling field-deployable testing. The incorporation of dual interrogation molecular assays and inhibitor-tolerant polymerases ensures high accuracy and robustness, even with challenging sample types. The protocol details the extraction process and subsequent RT-qPCR detection, facilitating efficient and reliable pathogen screening for plant diagnostics and disease management. In this workflow, we describe an extraction and probe-based RT-qPCR method used to detect viroid and virus genotypes in plant samples extracted from raw tissue or FTA imprinted cards.

Key words DNA, Amplicon, Reverse transcriptase, Preamplification, RT-qPCR , Hops latent viroid (HLVd), Beet Curly Top Virus (BCTV), Genotyping, Dual interrogation, Viroid, Virus, Plant diagnostics, FTA cards, Nucleic acid extraction

1 Introduction

Genotyping viruses and viroids is routinely carried out using reverse transcriptase quantitative PCR (qPCR) [1]. With good assay design and following standard validation execution, laboratories develop robust, precise, and accurate genotyping methods [2]. Diagnostic molecular probes can assess virus detection in various samples using a diagnostic approach to measure virus and viroid genotypic detection [3]. Coupling a reverse transcription step to polymerase chain reaction enables targeting of expressed virus genotypes as well as RNA viroids such as Hops latent viroid (HLVd) [4]. The accuracy and selectivity of the probe detection call can be expanded in approach for additional layers of confirmation by incorporating dual target interrogation where two separate detection events are carried out for viroids or viruses genotyping detection [5].

Anthony Torres (ed.), *Genotyping: Methods and Protocols*, Methods in Molecular Biology, vol. 2943,
https://doi.org/10.1007/978-1-0716-4642-7_10,

Molecular diagnostic specificity and accuracy in assay validation execution are essential to ensure the assay has robustness, reproducibility, accuracy, and specificity [2, 6]. Some probes may be designed for genotype presence or absence of variation or where specific SNPs are targeted for virus or viroid genotype detection [7, 8]. In our analysis, we developed dual interrogation multiplex viroid and virus detection assay by a modified reverse transcriptase with amplification enrichment and qPCR detection in *C. sativa* HLVd and BCTV genotype detection [5, 9]. A major pathogen of *C. sativa* is Hops latent viroid [10]. Ensuring a HLVd-free environment is a critical stage for nursery environments and good agricultural application and practice in this context [11]. Here, we present a high-throughput protocol for nucleic acid isolation from tissue or imprinted FTA card material, which is used as input to prepare reverse transcriptase and a preamplification-enriched quantitative PCR workflow.

2 Materials

2.1 Sample Preparation

1. Bunsen burner.
2. Calibrated p1, p100, p200 multichannel pipettes and p1, p200, p1000 single-channel pipettes.
3. 100–1000 μL nuclease-free aerosol filtered pipette tips.
4. 20–300 μL nuclease-free aerosol filtered pipette tips.
5. 1–100 μL nuclease-free aerosol filtered pipette tips.
6. 0.1–20 μL nuclease-free aerosol filtered pipette Tips.
7. Stainless steel dissecting forceps.
8. 250 mL beaker.
9. Optional: plate piercer.
10. 1.5 mL centrifuge tubes.
11. Gloves.
12. Isopropanol.
13. Nuclease-free water.
14. Extraction buffer solution: 0.25 M Tris–HCl, pH 8, 0.25% Triton-X.
15. FTA plant saver cards.
16. Unicore punch kit, 2 mm.

2.2 Reverse Transcriptase Preamplification Reaction

1. Standard PCR thermocycler.
2. Mini plate spinner.
3. Gloves.
4. Clean designated space.

5. Clear 96-well plate or similar.
6. Sterile loading trough.
7. RT-QuantiNova Probe qPCR Kit (Qiagen).
8. Genotype-specific positive DNA for the positive control template.
9. Virus/Viroid genotype-specific QPCR and internal positive control Oligonucleotides (*see* **Note 1**).
10. Virus/Viroid genotype-specific QPCR positive DNA for positive control template (*see* **Note 2**).
11. AlumaSeal II sealing foils.

2.3 qPCR Reaction

1. Standard real-time PCR thermocycler.
2. QuantiNova Probe qPCR Kit (Qiagen).
3. Genotype-specific positive DNA for the positive control template.
4. Virus/Viroid genotype-specific QPCR and internal positive control Oligonucleotides.
5. Virus/Viroid genotype-specific QPCR positive DNA for positive control template.
6. Virus/Viroid genotype-specific custom qPCR probes (*see* **Note 3**).
7. Optical sealing foils.

3 Methods

3.1 Sample Preparation

1. Prepare the workstation on the lab bench by wiping down the surface with alcohol where the preparation will be performed.
2. Prepare clean PCR setup with a clean 96-well plate for sample preparation in an easily accessible area.
3. Sterilize forceps by dipping them in alcohol and flame sterilizing them or aseptically wiping excess alcohol until all alcohol has evaporated.
4. Remove a small sampling of raw tissue pieces or imprinted FTA card (~2 mm) with flame-sterilized forceps or alcohol-sterilized QIAGEN 2 mm UniCore Punch tool, taking care not to touch the sample directly with gloves.
5. Carefully place the sample into the PCR plate well at the bottom of the well so that no contact is made with the top of the PCR plate.
6. Repeat **steps 3.1.3–3.1.5** with remaining lots/replicates until all samples are processed.

7. Fill sterile liquid basin/trough with Extraction Buffer using dedicated single channel or multichannel 30–300 μL multichannel pipettor loaded with clean tips to draw and dispense 100 μL of extraction buffer into each well.
8. Repeat for all remaining samples and replace tips each time. After, seal with AlumaSeal II film strip.
9. Spin plates down in plate spinner for 20 s to ensure all material is fully immersed in the extraction buffer.
10. Load the plate onto the thermocycler and set it to run for 5 min. Boil with the lid closed and the heated cover on.
    (a) ~95–100 °C 5 min 1×.
11. Prepare a clean 96-well plate(s) for performing a serial dilution with the boiled sample.
12. Following boil, fill ~10 mL of nuclease-free water into a new sterile liquid basin/trough and pipette the appropriate dilution volume into a clean PCR plate for a 1:100 dilution (e.g., 90 μL Water + 10 μL boiled sample serial diluted with 1:10 10 μL boiled diluted sample + 90 μL Water for 1:100 dilution) (*see* **Note 4**).
13. Spin plates down in plate spinner for 20 s to ensure all material is fully immersed in the extraction buffer.
14. Using a plate piercer or pipette tips, pierce the boiled extraction plate.
15. Perform a 1:100 dilution in dilution PCR plate(s) with the boiled sample.

3.2 Reverse Transcription Preamplification Enrichment Preparation

1. Prepare a RT-PCR setup with a clean 96-well plate for sample preparation in an easily accessible area.
2. In a dedicated pre-PCR space prepare Reverse Transcriptase PCR Master Mix according to following table:

Reagent	Volume for 1 reaction (μL)
QuantNova Probe RT-qPCR Master Mix 2×	7.5 (*see* **Note 5**)
100× QuantiNova RT Mix (Reverse transcriptase Taq Mix)	0.2
Nuclease-free water	1.5
Virus/viroid dual interrogation multiplex primer mix	1

3. Prepare RT PreAmp master mix in a sterile multichannel loading trough.
4. Set up a clean clear 96-well plate.
5. In the dedicated PCR hood, use a 0.1–10 μL multichannel pipette, load four to eight 0.1–20 μL filtered pipette tips, pipette 10 μL of the preAmp master mix to each column of the plate, respectively, for columns 1–11.
6. Eject tips into waste container.
7. For all reactions, using a 0.1–10 μL multichannel or single-channel pipette, draw and dispense 5 μL of 1:100 diluted crude extract from crude extraction plate to corresponding wells of RT PreAmp plate and eject tips after dispensing.

   NOTE: Ensure tips are below the surface of the preamplification master mix before dispensing.
8. Using an applicator, seal the plate with AlumaSeal II plate sealing film foil (sticky side down) and ensure that all wells are sealed.
9. Spin down in a mini plate spinner for 10 s.
10. Place the RT PreAmp plate in the thermocycler and select and run the RT PreAmp Enrichment PCR protocol:
    (a) 50 °C for 10 min 1×.
    (b) 95 °C for 5 min 1×.
    (c) 95 °C for 15 s.
    (d) 60 °C for 60 s (*see* **Note 6**).
    (e) Repeat **steps b–d** 10×.
    (f) 4 °C hold.
11. Following the preamplification reaction, fill ~10 mL of nuclease-free water into a new sterile liquid basin/trough.
12. Spin preamplification plates down in plate spinner for 20 s.
13. Using a plate piercer or pipette tips, pierce the wells of the preamplification enrichment plate.
14. Pipette 100 μL water to dilute the preamplification enrichment reaction to a 1:5 working dilution.

3.3 qPCR Virus/Viroid Genotyping Reaction Setup

1. Prepare a RT-PCR setup with a clean 96-well plate for sample preparation in an easily accessible area.
2. In a dedicated viroid/virus dual interrogation qPCR master mix according to the following:

Reagent	Volume for 1 reaction (μL)
QuantNova probe qPCR Master Mix 2×	3.75 (*see* **Note 5**)

(continued)

Reagent	Volume for 1 reaction (μL)
QPCR Probes (each)	0.2
Nuclease-free water	0.2
Dual interrogation multiplex assay primer mix	1

3. In the dedicated PCR space, use a 0.1–10 μL multichannel pipette, load 0.1–20 μL filtered pipette tips, pipette 5 μL of the qPCR reaction mix to each well of the plate, respectively, for each sample.
4. Change tips and prepare to load samples, use fresh tips for each new sample (*see* **Note 7**).
5. For all PreAmp enrichment reactions, using a 0.1–10 μL multichannel pipette or single-channel pipette, draw and dispense 5 μL of 1:100 diluted preamp solution to corresponding wells of qPCR plate, ensure the tips are below surface of mix before dispensing, mix by pipetting up and down several times, then, eject tips after final mix.
6. For all reactions with purified nucleic acid, using a 0.1–10 μL multichannel pipette or single-channel pipette, draw and dispense 5 μL of kit (1–10 ng/μL) purified nucleic acid from commercial kit or similar be sure tips are below surface of mix before dispensing, mix by pipetting up and down several times, then, eject tips after final mix.
7. Pipette to mix 1–3 times and eject tip.
8. To minimize bubbles in the qPCR reaction plate, pipette preamped the solution slowly.
9. After each sample well has been filled, obtain the control standards and aliquot 5 μL of positive control template, negative control template, and no template control into their each own appropriate wells.
10. Seal plate thoroughly with qPCR sealing foil.
11. Spin plates for 10 s.
12. The plate is now ready for data acquisition.
13. Place the qPCR plate in the real-time instrument and select and run the QPCR/multiplex protocol:
    (a) 95 °C for 3 min 1×.
    (b) 95 °C for 15 s.
    (c) 60 °C for 60 s (With Plate Read/Data Acquisition) (*see* **Note 6**).
    (d) Repeat **steps b** and **c** 40×.
14. Run the qPCR Reaction on the qPCR instrument with appropriate data acquisition settings.

3.4 qPCR Instrument Setup

3.4.1 Instrument Run: Qiagen ROTOR-GENE Q

1. Double Click Application Icon to Open.
2. To Create Protocol: Select New Protocol → Select Quantinova PCR → Enter Tube and Reaction Size → Enter PCR parameters as outlined in **step 3.3.13**, ensuring plate read step at appropriate annealing step for Assay (Table 1). Save protocol file → Click Run.
3. Enter all sample names as unknowns on the following screen and mark the PTC and NTC samples as such.
4. After the run is complete, click the analyze button and select the color channel that you intend to analyze.
5. Save the reaction as the qPCR Experiment barcode.
6. Example File Name Template: yyyymmdd_[Location] QPCRID###_initials_qPCR NOTE: Set a threshold for calling a Ct value and export the data to excel from the Rotor-gene software.
7. Once the run is complete, open Qiagen analysis software from the desktop.
8. Open datafile.
9. Select Analysis tab, select target color Cycling from directory for analysis (Green = FAM) (Orange = ROX), etc.
10. To all wells and apply the following analysis functions:
    (a) Dynamic tube.
    (b) Slope correction.

Table 1
Reverse Transcriptase (RT) enrichment and qPCR/multiplex amplification protocols with thermal cycling steps

Step	Temperature (°C)	Time	Repetitions	Notes
RT-pre-amp enrichment PCR protocol				
a	50	10 min	1×	
b	95	5 min	1×	
c	95	15 s	10×	
d	60	60 seconds		Plate read/data acquisition (*see* **Note 6**)
e	4	Hold	–	
qPCR/multiplex protocol				
a	95	3 min	1×	
b	95	15 s	40×	
c	60	60 s		Plate read/data acquisition (*see* **Note 6**)

The table details the temperature and duration of each step in the specified process, including a repeated cycle with data acquisition

11. Use auto threshold when able, however, if no standard curve is ran, set threshold between 0.03 and 0.05. Fluorescent target threshold can be adjusted manually at your supervisor's discretion during analysis; however in general, the threshold line should be set to start to mid exponential phase of amplification curve.
12. Once threshold is set, observe Ct values measured for each sample in the results summary window for each fluorescent color.
13. Confirm the presence or absence of signal in the controls and then in the sample wells both by looking at the shape for the second-derivative curves for each well and the software demarcation.
14. NOTE: Samples with a positive signal will have a numerical value after the Cp is calculated; negative signals will be blank.
15. Repeat for all targets and export table of Ct values to excel for each target.

3.4.2 Instrument Run and Analysis: Roche LightCycler

1. Double Click Application Icon to Open.
2. Open Light Cycler 480 software from the desktop.
3. Enter Username: Admin.
4. Password: Roche480.
5. Open qPCR experiment file.
6. In Experiment Tab, below Overview, select Apply Template, and follow directory: From System Admin → Templates → Run Templates. Select QPCR Test Protocol and click check mark.
7. In Sample Editor Tab, below plate grid, select apply template, and follow directory: From System admin → Templates → Sample Templates. Select QPCR Test Format and click check mark.
8. In Sample Editor Tab, in the lower right corner, select Import to upload the instrument file from LIMS. From the drop-down menus, select "Pos" for the left column and "Sample Name" for the right column.
9. Ensure that Rel Quant Protocol is selected in step one, and both FAM (465–510 nm) and HEX (533–580 nm) excitation filters are selected. Apply Cy5 and ROX channels additionally.
10. Save the reaction as the qPCR Experiment barcode:

    Example File Name Template: yyyymmdd_[Location] QPCRID###_initials_qPCR.
11. Select Start.
12. NOTE: If a window comes up requesting "Reasons for a Change," enter your initials and select the check mark.

13. Once the run is complete, click the Analysis tab and select the Absolute Quantification/Second Derivative Max analysis mode.
14. At the bottom of the page, click on the Filter tab and change the filter to HEX.
15. Select Calculate to display the TC/Reference Cps for the sample wells and for the controls.
16. Confirm the presence or absence of signal in the controls and then in the sample wells both by looking at the shape for the second-derivative curves for each well and the software demarcation.
17. Wells with detected signal will be marked red and those with non-detectable signals will be green. If the software is unable to discern whether the signal was detected, the well will be colored blue.
18. During analysis, samples observed with a positive signal will have a numerical value after the Cp is calculated; negative signals will be blank.
19. In cases of unexpected non-detects for the Reference Cp: Confirm the assessment by looking at the second-derivative curve. If the curve is consistent with that of a non-detect, record the sample name in your lab notebook as the sample will need to be re-prepped.
20. Refer to notes for the re-prep procedure.
21. In cases where the software is unable to make a call: Select the Drop-down menu and change the analysis type from High Confidence to High Sensitivity. If the well color of the uncertain wells becomes either red or green, record the value in lab notebook.
22. Confirm the new call by looking at the second derivative curve (s).
23. Repeat steps with the FAM filter that will display the Target Cps.

3.4.3 Instrument Run and Analysis: Quant Studio 5

1. Tap touchscreen of instrument to wake.
2. Login to Cloud Account.
3. Enter Username:
4. Enter Password:
5. Load experiment template from Table 1.
6. Select QPCR Test Method.
7. In Experiment Properties tab, enter experiment file as:

   Example yyyymmdd_[location] QPCRID###_initials_qPCR.

8. Ensure that File is Saved to Cloud.
9. Ensure that both FAM (465–510 nm) and VIC/HEX (533–580 nm) excitation filters are selected. Apply Cy5, ROX, and TAMRA channels additionally.
10. Open tray.
11. Load 96-well Plate.
12. Select Start.
13. Once the run is complete, click the Transfer File tab on the instrument to save archive data.
14. On the computer, open and browser and login to Thermo-Fisher Cloud Account: thermofisher.com/mysso/loginDisplay.
15. Select App Display: https://apps.thermofisher.com/apps/spa/#/apps.
16. Select and Open the Design and Analysis Application.
17. Select Summary, then Multicomponent plot to view target results (Fig. 1).
18. Select Open Existing Experiment, then search for and open: .eds experiment file.
19. In the experiment manager, select Export tab, ensure that results are selected, then export results and download the excel file.
20. The file is downloaded as a zipped folder, unzip the folder and locate the QPCR test results excel file.
21. In the excel home page, click on the Filter tab to enable filter function. Then, change the Target filter to Target 2 (VIC/HEX).
22. Confirm the presence or absence of signal in the controls and then in the sample wells both by analyzing the Target 2 (VIC/HEX) Cp data.
23. View the amplification curves in the Results Tab in the Design and Analysis application looking at the shape for the cycle amplification curves for each well and the software demarcation.
24. Each fluorescent target threshold can be adjusted manually at your supervisor's discretion during analysis; however in general, the threshold line should be set to start to mid exponential phase of amplification curve.
25. Wells with detected signal above threshold line will be marked with Cp value for Target 1 (FAM) and Target 2 (VIC/HEX) and those with non-detectable signals will be undetermined.

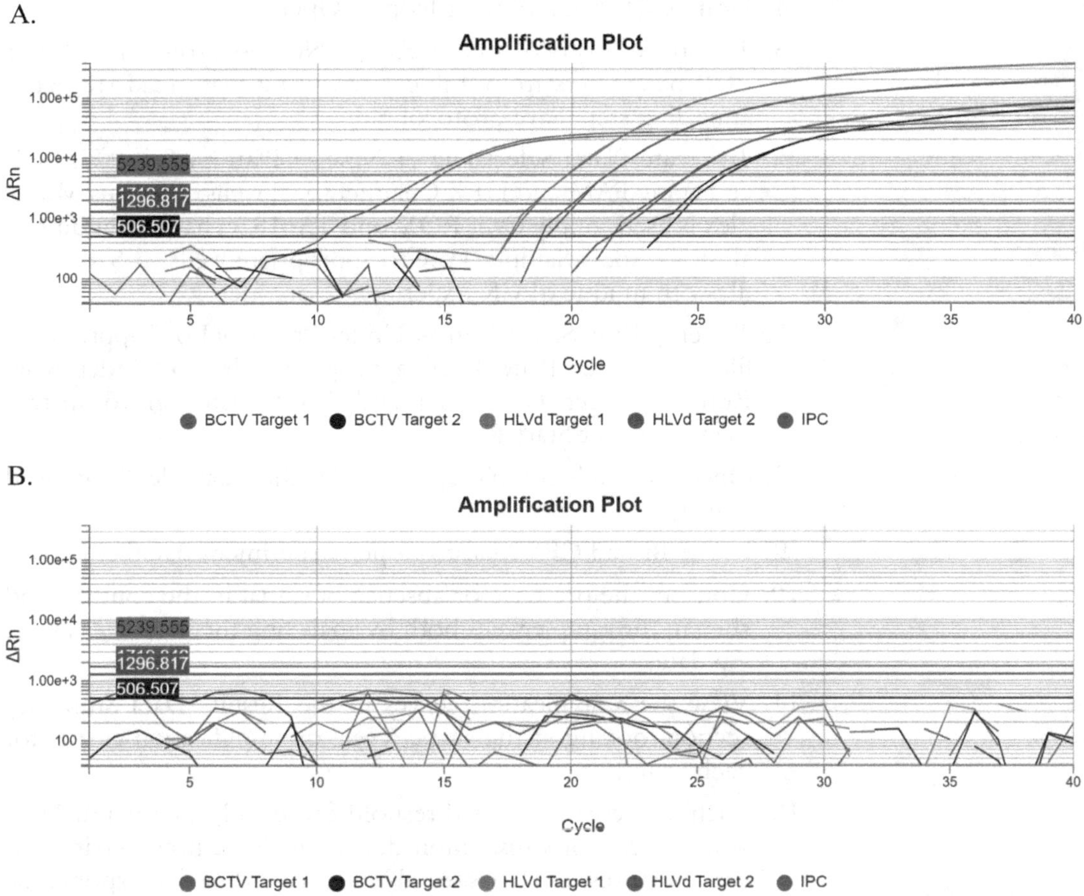

Fig. 1 Example amplification plots of observed fluorescent detection above threshold for dual interrogated HLVd viroid and BCTV virus genotypes in multiplex reaction samples (**a**) detection and (**b**) no detection

26. NOTE: Samples with a positive signal will have a numerical value for Target 1 (FAM) once the Cp is calculated, whereas negative signal will be undetermined.
27. In cases of unexpected non-detects or undetermined for the Target 2 (VIC/Hex): Confirm the assessment by looking at the amplification curve in the Results tab of the software.
28. If the curve is consistent with a non-detect result for all fluorophores, record the sample name in your lab notebook. The sample may need to be re-prepped to confirm the reaction failure.
29. Repeat steps with the FAM, Cy5, ROX, and TAMRA filters.

3.4.4 Instrument Run and Analysis: Biorad CFX96

1. Open Biorad CFX96 Maestro software from the desktop.
2. Enter Username:
3. Enter Password:

4. Double Click Instrument Icon to Open.
5. To Create Protocol: Select File → New → Protocol → Enter PCR parameters from Table 1, ensuring plate read step at X annealing step. Save protocol file.
6. To create Plate: Select File → New → Plate → Select all wells and apply FAM and HEX Channel to all wells. (If Virus Multiplex is used, Apply Cy5, ROX, and TAMRA channels additionally). → Apply Sample Type to appropriate wells → Save Plate file with unique qPCR Experiment barcode.
7. To setup Run: Select Run → Under Protocol Load appropriate file. → Under Plate Load appropriate file. → Under Start Run → Select Open Lid and Load plate on to instrument → select Start Run.
8. Once the run is complete, save and archive datafile if not saved already.
9. In the Biorad CFX Maestro, open experiment datafile.
10. Confirm the presence or absence of signal in the controls and then in the sample wells both by analyzing the Target 2 (HEX) Cp data.
11. View the amplification curves in the Results Data Summary section looking at the shape/cycle of amplification curves for each well.
12. Each fluorescent target threshold can be adjusted manually at your supervisor's discretion during analysis; however in general, the threshold line should be set to start to mid exponential phase of amplification curve.
13. Wells with detected signal above threshold line will be marked with Cp value for Target 1 (FAM) and Target 2 (HEX) and those with non-detectable signals will be undetermined.
14. NOTE: Samples with a positive signal will have a numerical value for Target 1 (FAM) once the Cp is calculated, whereas negative signal will be N/A.
15. In cases of unexpected non-detects or undetermined for the Target 2 (Hex):
16. Confirm the assessment by looking at the amplification curve in the results tab of the software.
17. If the curve is consistent with that of a non-detect, record the sample name in your lab notebook as the sample will need to be re-prepped.
18. Repeat **steps 3.3.3.11–3.4.4.10** with the Target 1 (FAM) filter that will display the Target 1 (FAM) Cps.

4 Notes

1. RT-qPCR oligonucleotides must be well designed to produce a single amplicon product generally not greater in length than 300 bp and compatible with thermodynamic properties of hybridization with a single-stranded fluorescently labeled oligonucleotide probe in a real-time RT-qPCR/qPCR reaction. Oligonucleotide primer should target conserved regions of the viroid or virus-specific gene or genotypes (e.g., Beet Curly top virus, Hops Latent Viroid, and Alfalfa mosaic virus).
2. Different genotypes that affect or have pathology implications in the model or target organism should be employed as positive controls. Positive control material can be isolated for recovery and subsequent testing. Additionally, synthetic products can be produced for viroid or virus genotype targets.
3. Ensure that the probe oligonucleotides are fluorescently labeled to contain a 5′ end with a unique spectral range fluorophore and a 3′ quencher. Designs for probes should be genotype target-specific to yield a presence/absence result for a specified genotype, do not possess cross-reactive overlap of an unintended target, and should be checked for homology. It is also important to select appropriate fluorophores for respective qPCR real-time instrument.
4. When pipetting from the extraction plate of crude lysis of fast nucleic acid extraction, ensure to pull from clear supernatant pipetting below the meniscus.
5. This protocol recommends a specific reverse transcriptase PCR and qPCR master mix. However, other vendors' inhibitor-tolerant RT-qPCR/qPCR master mixes can be utilized.
6. The annealing temperature for the designed oligonucleotide prepared for the target PCR primer mix should be at least 60 ° C.
7. Bubbles in PCR plates can induce qPCR artifacts in data analysis and should be avoided/monitored.

Conflict of Interest

The authors have no conflicts of interest to declare that are relevant to the content of this chapter.

References

1. Singh RK, Gopala (2021) Innovative approaches in diagnosis and management of crop diseases, The mollicutes, vol 1. Apple Academic Press
2. Saponari M, Loconsole G, Liao HH, Jiang B, Savino V, Yokomi RK (2013) Validation of high-throughput real time polymerase chain reaction assays for simultaneous detection of invasive citrus pathogens. J Virol Methods 193(2):478–486. https://doi.org/10.1016/j.jviromet.2013.07.002
3. James D, Varga A, Pallas V, Candresse T (2006) Strategies for simultaneous detection of multiple plant viruses. Can J Plant Pathol 28(1):16–29. https://doi.org/10.1080/07060660609507267
4. Puchta H, Ramm K, Sanger HL (1988) The molecular structure of hop latent viroid (HLV), a new viroid occurring worldwide in hops. Nucleic Acids Res 16(10):4197–4216. https://doi.org/10.1093/nar/16.10.4197
5. Torres A, Pauli C, Sarmiento C, Zalewski C, Gaudino R (2024) Differential expression analysis of Cannabis sativa response to thermotherapy of hops latent viroid (HLVd) infection and clearance of the viroid in tissue culture micropropagation. bioRxiv. https://doi.org/10.1101/2024.04.06.588422
6. Fernandez IMA, Parungao M, Hollin J, Selimotic B, Farrar G, Seyler T, Anand A, Ahmad R (2023) A novel, precise and high-throughput technology for viroid detection in Cannabis (MFDetect(TM)). Viruses 15(7). https://doi.org/10.3390/v15071487
7. Schmidt J, Berghaus S, Blessing F, Herbeck H, Blessing J, Schierack P, Rodiger S, Roggenbuck D, Wenzel F (2022) Genotyping of familial Mediterranean fever gene (MEFV)-single nucleotide polymorphism—comparison of Nanopore with conventional Sanger sequencing. PLoS One 17(3):e0265622. https://doi.org/10.1371/journal.pone.0265622
8. Jarugula S, Wagstaff C, Mitra A, Crowder D, Gang D, Rayapati N (2023) First reports of Beet curly top virus, Citrus yellow vein-associated virus, and Hop latent viroid in industrial hemp (Cannabis sativa) in Washington State. Plant Dis 107. https://doi.org/10.1094/PDIS-12-22-2981-PDN
9. Pauli C, Gaudino R, Torres A, Zalewski C (2021) Methods and compositions for pathogen detection in plants. United States Patent
10. Adkar-Purushothama CR, Sano T, Perreault JP (2023) Hop latent viroid: a hidden threat to the cannabis industry. Viruses 15(3). https://doi.org/10.3390/v15030681
11. Miotti N, Passera A, Ratti C, Dall'Ara M, Casati P (2023) A guide to cannabis virology: from the virome investigation to the development of viral biotechnological tools. Viruses 15(7). https://doi.org/10.3390/v15071532

Chapter 11

Haplotype and Presence–Absence Variation Analysis Using High-Resolution Melting (HRM) Analysis for Genotyping Genes of Interest

Anthony Torres, Robert Givens, and Reginald Gaudino

Abstract

This chapter outlines a basic protocol for DNA extraction, applicable to both FTA cards and leaf pieces. The extracted DNA then undergoes a pre-amplification enrichment reaction, followed by dilution and preparation for input into a real-time amplification with High-Resolution Melt (HRM) detection. The resulting melting genotypes are produced for comparison and analysis, providing a practical approach for genotyping studies. The flexibility of the method allows specific analytical output and high throughput analysis of a given DNA sequence that possesses inherent variability between individuals. This protocol describes the crude approach of extracting DNA from samples, preparing a real-time HRM reaction, and analyzing the result.

Key words DNA extraction, FTA cards, Real-time HRM, Genotyping, Pre-amplification, DNA, Amplicon, qPCR, HRM, Primers, Probe, Tm, Melting temp, Allele, Melting difference, Structural and allelic variations

1 Introduction

High-Resolution Melt (HRM) analysis is a real-time PCR method that indirectly detects/reveals genotype-specific variants in a targeted amplicon's base composition or sequence. The method can be performed to analyze genotypes of targeted regions from many different tissue types from various organisms [1–5]. A robust DNA extraction approach frees the nucleic acid to be analyzed by this routine molecular method. Well-designed, specific, targeted primers will determine the regions amplified and assessed under real-time PCR conditions. The technique captures the transition of DNA from double- to single-stranded and precisely measures, using an intercalating DNA dye and the re-hybridization of the double-stranded amplicon over small increments of temperature change coupled with fluorescent detection to determine the melt

Anthony Torres (ed.), *Genotyping: Methods and Protocols*, Methods in Molecular Biology, vol. 2943,
https://doi.org/10.1007/978-1-0716-4642-7_11,

temperature differential between amplicons [6]. Any interior sequence or structural variants that significantly alter the melting temperature of the amplicon after polymerase chain reaction (PCR) should concomitantly alter the respective High-Resolution Melt (HRM) profile of the amplicon as it rehybridizes [7].

As a genotype screening methodology, HRM offers a more expedient, cost-effective proxy for both sequencing or probe-based real-time PCR analysis [8]. HRM Analysis of DNA Melt profiles of PCR amplicons can clearly distinguish single or multiple nucleotide differences as well as structural variation in well-designed targeted regions in genes of interest that can be phenotypically or clinically significant [6, 9, 10]. It has a wide range of applications such as distinguishing closely related species [11], in agricultural biotechnology [6], as well as identifying mutation status in diagnostics [12]. In allelic-specific and genetic structure genotyping, insertions, deletions, or single nucleotide polymorphisms, high-resolution melt analysis of amplicons can reveal genotype-specific groupings based on genetic differences. When the genotype differences are observed in larger gene families or are between two different haplotypes, such as in the case of THCAS and CBDAS [13] or the detection of the Y chromosome in an XY plant system [6], HRM has been deployed as a powerful genotyping tool. This nucleic-acid-based analysis thus promises an expeditious, economical, robust saturation-point assay for prospective genotyping. We described a routine workflow and protocol for DNA extraction followed by a pre-enrichment and HRM preparation and analysis of targeted amplicons.

2 Materials

2.1 Sample Preparation

1. Bunsen burner.
2. Calibrated p1, p100, p200 multichannel pipettes and p1, p200, p1000 single-channel pipettes.
3. 100–1000 μL nuclease-free aerosol filtered pipette tips.
4. 20–300 μL nuclease-free aerosol filtered pipette tips.
5. 1–100 μL nuclease-free aerosol filtered pipette tips.
6. 0.1–20 μL nuclease-free aerosol filtered pipette tips.
7. Stainless steel dissecting forceps.
8. 250 mL beaker.
9. Optional: plate piercer.
10. 1.5 mL centrifuge tubes.
11. Gloves.
12. Isopropanol.

13. Nuclease-free water.
14. Extraction buffer solution: 0.25 M Tris–HCl, pH 8, 0.25% Triton-X.
15. FTA plant saver cards.
16. Unicore punch kit, 2 mm.

2.2 Pre-amplification Reaction

1. Standard PCR thermocycler.
2. Mini plate spinner.
3. Gloves.
4. Clean designated space.
5. Clear 96-well plate or similar.
6. Sterile loading trough.
7. Genotype-specific qPCR oligonucleotides (<300 bp targets) (*see* **Note 1**).
8. GoTaq PCR master mix (Promega).
9. Genotype-specific positive DNA for the positive control template.
10. AlumaSeal II sealing foils.

2.3 HRM Reaction

1. Standard real-time PCR thermocycler.
2. Genotype-specific oligonucleotides (<300 bp targets) (*see* **Note 1**).
3. Typeit II HRM master mix kit (Qiagen) (*see* **Note 7**).
4. Genotype-specific positive DNA for the positive control template (*see* **Note 2**).
5. Optical sealing foils.

3 Methods

3.1 Sample Preparation

1. Prepare the workstation on the lab bench by wiping down the surface with alcohol where the preparation will be performed.
2. Prepare clean PCR setup with a clean 96-well plate for sample preparation in an easily accessible area.
3. Sterilize forceps by dipping them in alcohol and flame sterilizing them or aseptically wiping excess alcohol until all alcohol has evaporated.
4. Remove a small sampling of raw tissue pieces or imprinted FTA Card (~2 mm) with flame-sterilized forceps or alcohol-sterilized QIAGEN 2 mm UniCore punch tool, taking care not to touch the sample directly with gloves.

5. Carefully place the sample into the PCR plate well at the bottom of the well so that no contact is made with the top of the PCR plate.
6. Repeat **steps 3–5** with remaining lots/replicates until all samples are processed.
7. Fill sterile liquid basin/trough with Extraction Buffer using dedicated single channel or multichannel 30–300 μL multichannel pipettor loaded with clean tips to draw and dispense 100 μL of extraction buffer into each well.
8. Repeat for all remaining samples and replace tips each time. After, seal with AlumaSeal II film strip.
9. Spin plates down in plate spinner for 20 s to ensure all material is fully immersed in the extraction buffer.
10. Load the plate onto the thermocycler and set it to run for 5 min. Boil with the lid closed and the heated cover on.
    (a) ~95–100 °C 5 min 1×.
11. Prepare a clean 96-well plate(s) for performing a serial dilution with the boiled sample.
12. Following boil, fill ~10 mL of nuclease-free water into a new sterile liquid basin/trough and pipette the appropriate dilution volume into a clean PCR plate for a 1:100 dilution (e.g., 90 μL water + 10 μL boiled sample serial diluted with 1:10 10 μL boiled diluted sample + 90 μL water for 1:100 dilution).
13. Spin plates down in plate spinner for 20 s to ensure all material is fully immersed in the extraction buffer.
14. Using a plate piercer or pipette tips, pierce the boiled extraction plate.

    Perform a 1:100 dilution in dilution PCR plate(s) with the boiled sample.

3.2 Pre-amplification Enrichment Reaction Setup

1. Prepare a clean PCR setup with a clean 96-well plate for sample preparation in an easily accessible area.
2. In a dedicated pre-PCR space, prepare a PCR master mix according to the following:

    Reagent:
    GoTaq PCR master mix: 12.5 μL.
    Genotype-specific Primer Mix (10 μM): 1 μL.
    Nuclease-free water: 7 μL.
3. Prepare PreAmp master mix in a sterile multichannel loading trough.
4. Set up a clean, clear 96-well plate.
5. In the dedicated PCR hood, use a 0.1–10 μL multichannel pipette, load four to eight 0.1–20 μL filtered pipette tips,

pipette 20 μL of the PreAmp master mix to each column of the plate, respectively, for columns 1–11.

6. Eject tips into waste container.
7. For all reactions, using a 0.1–10 μL multichannel or single-channel pipette, draw and dispense 5 μL of 1:100 diluted crude extract from crude extraction plate to corresponding wells of the pre-amplification plate and eject tips after dispensing (*see* **Note 3**).
8. Using an applicator, seal the plate with AlumaSeal II plate sealing film foil (sticky side down) and ensure that all wells are sealed.
9. Spin down in a mini plate spinner for 10 s.
10. Place the PreAmp plate in the thermocycler and select and run the PreAmp Enrichment PCR protocol:
    (a) 95 °C for 5 min 1×.
    (b) 95 °C for 15 s.
    (c) 60 °C for 30 s (*see* **Note 5**).
    (d) 72 °C for 30 s.
    (e) Repeat **steps a–d** 10×.
    (f) 4 °C hold.
11. Following the pre-amplification reaction, fill ~10 mL of nuclease-free water into a new sterile liquid basin/trough.
12. Spin pre-amplification plates down in plate spinner for 20 s.
13. Using a plate piercer or pipette tips, pierce the wells of the pre-amplification enrichment plate.
14. Pipette 100 μL water to dilute the pre-amplification enrichment reaction to a 1:5 working dilution.

3.3 HRM Reaction Setup

1. Prepare clean PCR setup designated space with a new 96-well plate for sample preparation in an easily accessible area.
2. Prepare a HRM master mix according to the following:
    Reagent:
    Typeit II HRM master mix 2×: 3.75 μL (*see* **Note 4**).
    Genotype-specific primer mix (10 μM): 1 μL.
    Nuclease-free water: 0.25 μL.
3. In the dedicated PCR space, use a 0.1–10 μL multichannel pipette, load 0.1–20 μL filtered pipette tips, pipette 5 μL of the HRM reaction mix to each well of the plate, respectively, for each sample.
4. Change tips and prepare to load samples, use fresh tips for each new sample.

(a) For all PreAmp enrichment reactions, using a 0.1–10 μL multichannel pipette or single-channel pipette, draw and dispense 5 μL of 1:5 diluted PreAmp solution to corresponding wells of HRM plate, ensure tips are below surface of mix before dispensing, mix by pipetting up and down several times, then, eject tips after final mix.

(b) For all reactions with purified genomic DNA, using a 0.1–10 μL multichannel pipette or single-channel pipette, draw and dispense 5 μL of kit (1–10 ng/μL)* purified genomic DNA from DNEasy or similar be sure tips are below the surface of mix before dispensing, mix by pipetting up and down several times, then, eject tips after final mix.

5. Pipette to mix 1–3 times and eject tip.
6. To minimize bubbles in HRM reaction plate, pipette pre-amped solution slowly (*see* **Note 6**).
7. After each sample well has been filled, obtain the control standards and aliquot 5 μL of positive control template, negative control template, and no template control into their own appropriate wells.
8. Seal the plate thoroughly with optical sealing foil.
9. Spin the plate for 10 s.
10. The plate is now ready for data acquisition.
11. Place the HRM plate in the real-time thermocycler instrument and select to run the HRM protocol:
    (a) 50 °C for 2 min 1×.
    (b) 95 °C for 10 min 1×.
    (c) 95 °C for 15 s.
    (d) 60 °C for 30 s (With Plate Read/Data Acquisition) (*see* **Note 5**).
    (e) 72 °C for 30 s.
    (f) Repeat **steps c–e** 40×.
    (g) 95 °C for 15 s.
    (h) 60 °C for 1 min, then ramp at 0.15 °C/s to 95 °C (with a plate read/data acquisition) (*see* **Note 7**).
12. Run the HRM reaction on a real-time thermocycler instrument with data acquisition settings (*see* **Note 7**).

3.4 HRM Instrument Setup

3.4.1 Instrument Run: Roche Lightcycler

1. Double Click Application to Open the Roche Lightcycler Software from the desktop.
2. Enter Username: Admin.
3. Password: Roche480.

4. Open HRM experiment file.
5. In Experiment Tab, below Overview, select Apply Template, and follow directory: From System Admin → Templates → Run Templates. Select HRM Test Protocol and click check mark.
6. In Sample Editor Tab, below plate grid, select apply template, and follow directory: From System admin → Templates → Sample Templates. Select HRM Test Format and click check mark.
7. In Sample Editor Tab, in the lower right corner, select Import to upload the instrument file from LIMS. From the drop-down menus, select "Pos" for the left column and "Sample Name" for the right column.
8. Ensure that HRM protocol is selected in step one with the appropriate genotype controls and FAM (465–510 nm) excitation filters are selected for HRM.
9. Save the reaction as the HRM Experiment barcode:
10. File Name Template: yyyymmdd_[Location] HRMID###_initials_HRM.
11. Select Start.
12. Upon changes, a window requesting "Reasons for a Change" may appear. Enter your initials and select the check mark.
13. Once the run is complete, click the Analysis tab and select the Melt Curve Genotyping analysis mode.
14. Compare Melt Curve Genotyping Profiles to Genomic Standards and adjust with New Calls as needed. (You can also adjust Pre and Post Melt Sliders if required (Fig. 1)).
15. Export Melt Curve Genotyping Table by right clicking in Results Table and save the Results Table text file.
16. Export Melt Curve Genotyping Profiles by right clicking in Melting peaks Plot and save Plot as JPEG.

3.4.2 Instrument Run: Quant Studio 5

1. Tap touchscreen of instrument to wake.
2. Login to Cloud Account.
3. Enter Username:
4. Enter Password:
5. Load experiment template as detailed in **step 3.3.11**.
6. Select HRM test method.
7. In Experiment Properties tab, enter experiment file as:
8. yyyymmdd_[location] HRMID###_initials_qPCR.
9. Ensure that File is saved to Cloud.
10. Ensure that both FAM (465–510 nm) excitation filter is selected.

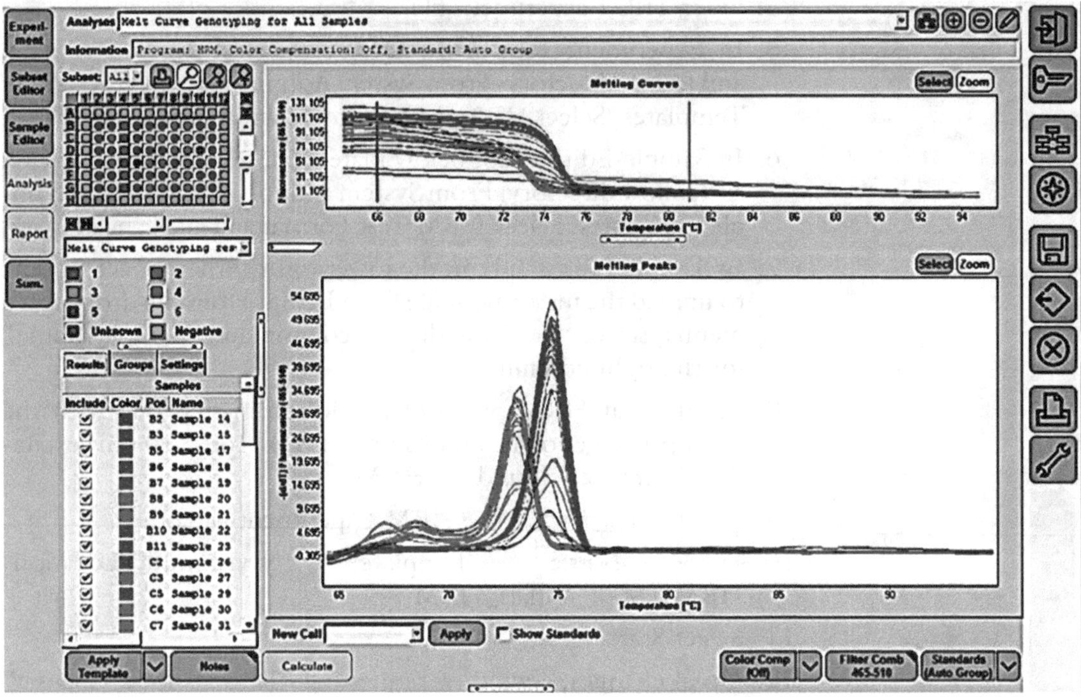

Fig. 1 Melt peak analysis, determined by melt curve sliders set at maximum and minimum fluorescence in the high-resolution melting acquisition, identified two major genotype groups. These groups are differentiated by their melting peak temperatures, with red peaks representing samples exhibiting a lower melting temperature relative to the blue peaks. Each distinct melting peak corresponds to a specific genotype/haplotype

11. Open tray.
12. Load 96-well plate.
13. Select start.
14. Once the run is complete, click the Transfer File tab on the instrument to save archive data.
15. On the computer, open and browser and login to Thermo-Fisher Cloud Account: thermofisher.com/mysso/loginDisplay.
16. Select App Display: https://apps.thermofisher.com/apps/spa/#/apps.
17. Select and Open the Design and Analysis Application.
18. Select Open Existing Experiment, then search for and open:
19. yyyymmdd_[location] HRMID###_initials_HRM.eds experiment file.
20. In the experiment manager, select Quality tab, ensure that Melt Curve (derivative) Plot are selected, then export results and download the excel file. Record Identified Genotype Melt Temperature peaks and Export Melt Curve Genotyping

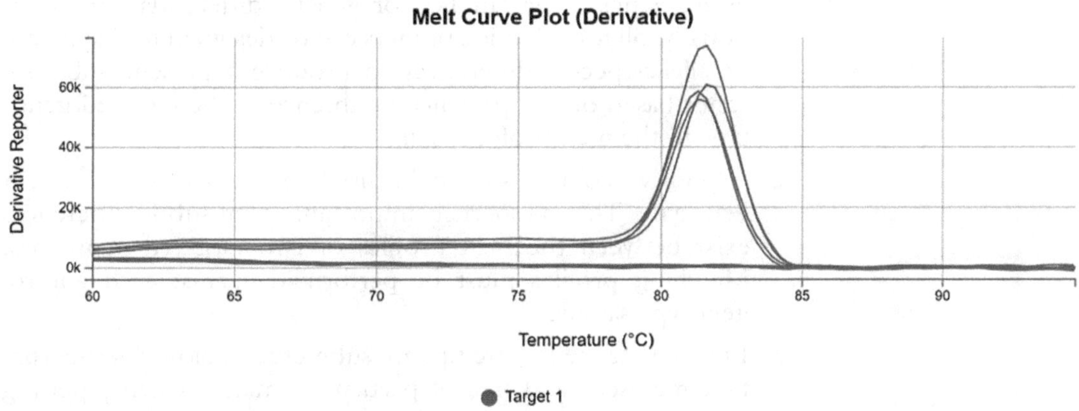

Fig. 2 Presence/absence determination of genotype by melt curve analysis. Samples present for targeted genotype observed fluorescence by derivative reporter

Profiles by right clicking in Melting peaks Plot and save Plot as JPEG (Fig. 2).

21. The file is downloaded as a zipped folder, unzip the folder and locate the HRM test results excel file.
22. In the excel home page, click on the Filter tab to enable filter function. Then, change the Target filter for Target 1 (FAM) to the appropriate Tm for selection.
23. Confirm the presence and Tm or absence of genotype signal in the controls and then in the sample wells both by analyzing the Target 1 (FAM) HRM Melt Data.
24. View the amplification curves in the Quality Summary Melt Curve Plot Tab in the Design and Analysis application looking at the shape and melt derivative curves for each well and the software demarcation.

4 Notes

1. Oligonucleotides used for high-resolution melting analysis polymerase chain reaction (HRM-PCR) should be specific and well designed. Target amplicon differences between genotypes should be significantly distinct in melting temperature (ideally, separation of melt profiles would be observed at a delta of ±1 °C or greater). When performing mutation analysis, primers should surround the targeted mutation. If the desired target is a structural difference between two genotypes, such as an insertion or deletion, oligonucleotides should surround the targeted structural difference. Generally, oligonucleotides should be specific to conserved and specifically designed to

target genes, gene families, or genetic differentiation. Additionally, oligonucleotide primers can be designed for haplotype or allelic-specific differences to produce a presence–absence result based on the presence or absence of the targeted genotype in the melt profile result.

2. A positive control should be used for each observed target genotype. This is especially important when subtle differences exist between the melt profiles of two targeted genotypes. Matching profiles must be performed against each control genotype sample.
3. Ensure that the pipette tips are submerged below the supernatant meniscus and free of particulate matter when pipetting from fast nucleic acid crude lysate because excess particulate matter can inhibit PCR.
4. The ideal annealing temperature for the workflow is 60 °C; however, this may vary depending on the primer design of the amplicon target.
5. This protocol uses the Typeit II HRM master mix from Qiagen, but other similar robust and inhibitor-tolerant HRM master mixes can be utilized with 465 nm emitting DNA intercalating dye.
6. Small bubbles in the HRM reaction can interfere with fluorescence capture of amplification or melting analysis. Ensure no bubbles are present in the reaction using a pipette tip before sealing the plate or by spinning a sealed plate down in a PCR plate centrifuge.
7. The melt curve parameters may vary by the master mix used. Use the recommended melt curve ramp and rate for your specific master mix. To ensure appropriate genotype results, the melt curve analysis is performed with sufficient data captured during melt curve acquisition.

Conflict of Interest

The authors have no conflicts of interest to declare that are relevant to the content of this chapter.

References

1. Simko I (2016) High-resolution DNA melting analysis in plant research. Trends Plant Sci 21(6):528–537. https://doi.org/10.1016/j.tplants.2016.01.004
2. Hjelmso MH, Hansen LH, Baelum J, Feld L, Holben WE, Jacobsen CS (2014) High-resolution melt analysis for rapid comparison of bacterial community compositions. Appl Environ Microbiol 80(12):3568–3575. https://doi.org/10.1128/AEM.03923-13
3. Landolt P, Stephan R, Scherrer S (2019) Development of a new High Resolution Melting (HRM) assay for identification and differentiation of Mycobacterium tuberculosis

complex samples. Sci Rep 9(1):1850. https://doi.org/10.1038/s41598-018-38243-6

4. Ngui R, Lim YA, Chua KH (2012) Rapid detection and identification of human hookworm infections through high resolution melting (HRM) analysis. PLoS One 7(7):e41996. https://doi.org/10.1371/journal.pone.0041996
5. Robertson T, Bibby S, O'Rourke D, Belfiore T, Agnew-Crumpton R, Noormohammadi AH (2010) Identification of Chlamydial species in crocodiles and chickens by PCR-HRM curve analysis. Vet Microbiol 145(3–4):373–379. https://doi.org/10.1016/j.vetmic.2010.04.007
6. Torres A, Pauli C, Givens R, Argyris J, Allen K, Monfort A, Gaudino RJ (2022) High-throughput methods to identify male Cannabis sativa using various genotyping methods. J Cannabis Res 4(1):57. https://doi.org/10.1186/s42238-022-00164-7
7. Cousins MM, Donnell D, Eshleman SH (2013) Impact of mutation type and amplicon characteristics on genetic diversity measures generated using a high-resolution melting diversity assay. J Mol Diagn 15(1):130–137. https://doi.org/10.1016/j.jmoldx.2012.08.008
8. Montgomery JL, Sanford LN, Wittwer CT (2010) High-resolution DNA melting analysis in clinical research and diagnostics. Expert Rev Mol Diagn 10(2):219–240. https://doi.org/10.1586/erm.09.84
9. Krypuy M, Newnham GM, Thomas DM, Conron M, Dobrovic A (2006) High resolution melting analysis for the rapid and sensitive detection of mutations in clinical samples: KRAS codon 12 and 13 mutations in non-small cell lung cancer. BMC Cancer 6:295. https://doi.org/10.1186/1471-2407-6-295
10. Chatzidimopoulos M, Ganopoulos I, Madesis P, Vellios E, Tsaftaris A, Pappas AC (2014) High-resolution melting analysis for rapid detection and characterization of Botrytis cinerea phenotypes resistant to fenhexamid and boscalid. Plant Pathol 63(6):1336–1343. https://doi.org/10.1111/ppa.12210
11. Xanthopoulou A, Ganopoulos I, Tryfinopoulou P, Panagou EZ, Osanthanunkul M, Madesis P, Kizis D (2019) Rapid and accurate identification of black aspergilli from grapes using high-resolution melting (HRM) analysis. J Sci Food Agric 99(1):309–314. https://doi.org/10.1002/jsfa.9189
12. Ju M, Bai X, Zhang T, Lin Y, Yang L, Zhou H, Chang X, Guan S, Ren X, Li K, Wang Y, Li G (2020) Mutation spectrum of COL1A1/COL1A2 screening by high-resolution melting analysis of Chinese patients with osteogenesis imperfecta. J Bone Miner Metab 38(2):188–197. https://doi.org/10.1007/s00774-019-01039-3
13. Borna T, Salami SA, Shokrpour M (2017) High resolution melting curve analysis revealed SNPs in major cannabinoid genes associated with drug and non-drug types of cannabis. Biotechnol Biotechnol Equip 31:1–7. https://doi.org/10.1080/13102818.2017.1333456

Chapter 12

PCR Allele Competitive Extension (PACE) Workflow for Genotyping by SNP Identification

Anthony Torres, Jason Argyris, and Reginald Gaudino

Abstract

This chapter presents protocols for advanced genotyping and genetic analysis in plant science, particularly utilizing DNA extracted from FTA samples and leaf tissue. PACE (PCR Allele Competitive Extension) enables high-throughput SNP genotyping with remarkable accuracy, specifically targeting and differentiating between two alleles of interest. The protocol involves using allele-specific primers and a PACE master mix for selective extension and labeling, culminating in the clear identification of SNP genotypes. This method provides a valuable tool for SNP genotyping, efficiently detecting specific and targeted genetic variations associated with desired traits. Here, we describe a simplified workflow for SNP genotyping using this approach to detect SNPs in target genes of a dioecious plant.

Key words PACE, SNP genotyping, Allele-specific primers

1 Introduction

The ability to accurately and efficiently genotype Single Nucleotide Polymorphisms (SNPs) has become essential for unraveling the complex relationship between genotype and phenotype [1–3]. In this pursuit, PCR Allele Competitive Extension (PACE) is a robust and cost-effective method, offering distinct advantages over traditional SNP genotyping techniques [4]. The principle of PACE is rooted in the utilization of allele-specific primers designed to perfectly complement one SNP allele while intentionally mismatching with the alternative allele [5]. This specificity, coupled with the action of a thermostable polymerase, results in the selective extension of the perfectly matched primer and incorporation of a labeled nucleotide [6, 7]. Subsequent detection of this label allows for the accurate and unambiguous identification of the SNP genotype [8].

The applications of PACE are particularly impactful in the realm of plant improvement and agriculture [8, 9]. PACE enables the rapid and cost-effective genotyping of large plant populations,

Anthony Torres (ed.), *Genotyping: Methods and Protocols*, Methods in Molecular Biology, vol. 2943, https://doi.org/10.1007/978-1-0716-4642-7_12,

facilitating a range of critical activities [4]. These include in-depth genetic studies to understand the basis of valuable traits, marker-assisted selection to accelerate breeding programs and develop improved cultivars, and an understanding of genotype-to-phenotype correlation in gene characterization studies [10, 11]. Its versatility and broad applicability across various fields demonstrate the importance of the technique for understanding the impact of genetic variation on diverse phenotypes, such as flowering time and photoperiodism in *C. sativa* [12–14]. Here, we describe a protocol for preparing PACE SNP genotyping reactions from FTA-prepared DNA extracts that can be used to classify SNPs from individuals while developing a genotyping database of tested plant accessions.

2 Materials

2.1 Sample Preparation

1. Bunsen burner.
2. Calibrated p1, p100, p200 multichannel pipettes and p1, p200, p1000 single-channel pipettes.
3. 100–1000 μL nuclease-free aerosol filtered pipette tips.
4. 20–300 μL nuclease-free aerosol filtered pipette tips.
5. 1–100 μL nuclease-free aerosol filtered pipette tips.
6. 0.1–20 μL nuclease-free aerosol filtered pipette tips.
7. Stainless steel dissecting forceps.
8. 250 mL beaker.
9. Optional: plate piercer.
10. 1.5 mL centrifuge tubes.
11. Gloves.
12. Isopropanol.
13. Nuclease-free water.
14. Extraction buffer solution: 0.25 M Tris–HCl, pH 8, 0.25% Triton-X.
15. Salt buffer solution: 5 M NaCl.
16. FTA plant saver cards.
17. Unicore punch kit, 2 mm.

2.2 PACE Reaction

1. Standard real-time PCR thermocycler.
2. Allele-specific FRET Cassette Nested PACE Oligonucleotides (<300 bp targets) (*see* **Note 1**).
3. Allele-specific PACE primer mix (*see* **Note 2**).
4. PACE2.0 Genotyping master mix (3CRBioSciences).

5. Genotype-specific positive DNA for the positive control template.
6. Optical sealing foils.

3 Methods

3.1 Sample Preparation

1. Prepare the workstation on the lab bench by wiping the surface with alcohol where the preparation will be performed.
2. Prepare a clean PCR setup with clean 1.5 mL microcentrifuge tubes in an easily accessible area for sample preparation.
3. Sterilize forceps by dipping them in alcohol and flame sterilizing them or aseptically wiping excess alcohol until all alcohol has evaporated.
4. Remove a small sampling of raw tissue pieces or imprinted FTA Card (~2–8 mm) with flame-sterilized forceps or alcohol-sterilized QIAGEN 2 mm UniCore punch tool, taking care not to touch the sample directly with gloves (*see* **Note 3**).
5. Carefully place the sample into a 1.5 mL microcentrifuge tube at the bottom of the well so that it does not contact the top of the tube.
6. Repeat **steps 2–5** with remaining lots/replicates until all samples are processed.
7. Fill the sterile liquid basin/trough with Extraction Buffer Solution using a dedicated single-channel or multichannel 30–300 μL multichannel pipettor loaded with clean tips to draw and dispense 200 μL of extraction buffer into each tube.
8. Repeat for all remaining samples, replacing tips and ensuring tubes are capped between steps.
9. Spin tubes down in microcentrifuge for 60 s to fully immerse all material in the extraction buffer.
10. Load the tube on a thermal incubation block and set to run for 10 min. Boil with lids sealed with tube lock to close.
    (a) ~95–100 °C 10 min 1×.
11. Following boil, add 8 μL 5 M NaCl and vortex.
12. Add 400 μL 100% ethanol and invert several times.
13. Place tubes at −80 °C overnight for 15 min or −20 °C to precipitate DNA.
14. Following ultra-cold precipitation, mark the position where the pellet will form on the tube and centrifuge 1.5 mL tubes for 30 min at 4 °C.
15. Remove tubes carefully, decant the supernatant, and retain the pellet.

(a) Optional: Perform a 70% ethanol wash to remove additional residual waste.

16. Following wash, centrifuge 1.5 mL tubes at 12,000 × *g* or greater for 30 min at 4 °C.
17. Remove tubes carefully, decant the supernatant, and retain the pellet.
18. Allow the pellets to dry for 5–10 min at room temperature (*see* **Note 4**), then reconstitute the pellet in 60 μL of nuclease-free water.

3.2 PACE PCR Reaction Setup

1. Prepare a clean 1.5 mL centrifuge tubes for master mix and clean 96-Well PCR plate for targeted long range PCR reactions.
2. Prepare a targeted allele-specific PACE PCR reactions with 10 μL final reactions with PACE2.0 Genotyping master mix and custom allele-specific PACE primer mix for the target SNP of interest (*see* **Note 5**) according to below:

   Reagent:
   2× PACE2.0 Genotyping Master Mix: 5 μL.
   Allele-Specific PACE Primer Mix (10 μM): 0.138 μL.
3. In the dedicated PCR space, use a 0.1–10 μL multichannel pipette, load 0.1–20 μL filtered pipette tips, and pipette 5 μL of the prepared reaction mix to each well of the plate, respectively, for each sample.
4. Change tips and prepare to load samples, use fresh tips for each new sample.

   (a) For all reactions tested with reconstituted DNA input, using a 0.1–10 μL multichannel pipette or single-channel pipette, draw and dispense 5 μL of reconstituted DNA input to ensure that tips are below surface of mix before dispensing, mix by pipetting up and down several times, then, eject tips after final mix.

   (b) For all reactions with purified genomic DNA, using a 0.1–10 μL multichannel pipette or single-channel pipette, draw and dispense 5 μL of kit (10 pg–1 μg/μL) purified genomic DNA from DNEasy or similar, ensure that tips are below surface of mix before dispensing, mix by pipetting up and down several times, then, eject tips after final mix.
5. After each sample well has been filled, obtain the control standards and aliquot 5 μL of positive control template, negative control template, and no template control into their respective wells.
6. Seal plate thoroughly with optical qPCR sealing foil.

7. Spin plates for 10 s.
8. The plate is now ready for PACE PCR.
9. Place the PCR plate in the thermocycler instrument and select and run the PACE PCR protocol:
   - (a) 94 °C for 15 min 1×.
   - (b) 94 °C for 20 s.
   - (c) 65 °C for 60 s (dropping 0.8 °C each cycle to 57 °C) (*see* **Note 6**).
   - (d) Repeat **steps b** and **c** 10×.
   - (e) 94 °C for 20 s.
   - (f) 57 °C for 60 s.
   - (g) Repeat **steps e** and **f** 35×.
   - (h) 4 °C hold
10. Run PACE reaction on qPCR instrument with appropriate data acquisition settings (*see* **Note 7**).

3.3 qPCR Instrument Setup

3.3.1 Instrument Run: Qiagen ROTOR-GENE Q

1. Double Click Application Icon to Open.
2. To Create Protocol: Select New Protocol → Select Quantinova PCR → Enter Tube and Reaction Size → Enter PCR parameters from **step 3.2.9**, ensuring plate read step at appropriate annealing step for Assay (Table 1). Save protocol file → Click Run.
3. Enter all sample names as unknowns on the following screen and mark the PTC and NTC samples as such.

Table 1
PCR Allele Competitive Extension (PACE) amplification protocol and thermal cycling steps

Step	Temperature (°C)	Time	Repetitions	Notes
PACE PCR protocol				
a	94	15 min	1×	
b	94	20 s	10×	
c	65–57	60 s		Decrease of 0.8 °C each cycle repetition to 57 °C (*see* **Note 6**)
d	94	20 s	35×	
e	57	60 s		Plate read/data acquisition (*see* **Note 7**)
f	4	Hold	–	

The table details the temperature and duration of each step in the specified process, including a repeated cycle with data acquisition

4. After the run is complete, click the analyze button and select the color channel that you intend to analyze.
5. Save the reaction as the qPCR Experiment barcode.
6. Example File Name Template: yyyymmdd_[Location] QPCRID###_initials_qPCR NOTE: Set a threshold for calling a Ct value and export the data to excel from the Rotor-gene software.
7. Once the run is complete, open Qiagen analysis software from the desktop.
8. Open datafile.
9. Select Analysis tab, select target color Cycling from directory for analysis (Green = FAM) (Orange = ROX) etc.
10. To all wells and apply the following analysis functions:
11. Dynamic tube.
12. Slope correction.
13. Use auto threshold when able, however, if no standard curve is ran, set threshold between 0.03 and 0.05. Fluorescent target threshold can be adjusted manually at your supervisor's discretion during analysis; however, in general the threshold line should be set to start to mid exponential phase of amplification curve.
14. Once threshold is set, observe Ct values measured for each sample in the results summary window for each fluorescent color.
15. Confirm the presence or absence of signal in the controls and then in the sample wells both by looking at the shape for the second-derivative curves for each well and the software demarcation.
16. NOTE: Samples with a positive signal will have a numerical value after the Cp is calculated; negative signals will be blank.
17. Repeat for all targets and export table of Ct values to excel for each target.

3.3.2 Instrument Run and Analysis: Roche LightCycler

1. Double Click Application to Open.
2. Open LightCycler 480 software from the desktop.
3. Enter Username: Admin.
4. Password: Roche480.
5. Open qPCR experiment file.
6. In Experiment Tab, below Overview, select Apply Template and follow directory: From System Admin → Templates → Run Templates. Select QPCR Test Protocol and click check mark.

7. In Sample Editor Tab, below plate grid, select apply template and follow directory: From System admin → Templates → Sample Templates. Select QPCR Test Format and click check mark.
8. In Sample Editor Tab, in the lower right corner, select Import to upload the instrument file from LIMS. From the drop-down menus, select "Pos" for the left column and "Sample Name" for the right column.
9. Ensure that Rel Quant Protocol is selected in step one, and both FAM (465–510 nm) and HEX (533–580 nm) excitation filters are selected.
10. Save the reaction as the qPCR Experiment barcode:

    Example File Name Template: yyyymmdd_[Location] QPCRID###_initials_qPCR.
11. Select Start.
12. NOTE: If a window comes up requesting "Reasons for a Change," enter your initials and select the check mark.
13. Once the run is complete, click the Analysis tab and select the Absolute Quantification/Second-Derivative Max analysis mode.
14. At the bottom of the page, click on the Filter tab and change the filter to HEX.
15. Select Calculate to display the TC/Reference Cps for the sample wells and for the controls.
16. Confirm the presence or absence of signal in the controls and then in the sample wells both by looking at the shape for the second-derivative curves for each well and the software demarcation.
17. Wells with detected signal will be marked red and those with non-detectable signals will be green. If the software is unable to discern whether the signal was detected, the well will be colored blue.
18. During analysis, samples observed with a positive signal will have a numerical value after the Cp is calculated; negative signals will be blank.
19. In cases of unexpected non-detects for the Reference Cp: Confirm the assessment by looking at the second-derivative curve. If the curve is consistent with that of a non-detect, record the sample name in your lab notebook as the sample will need to be re-prepped.
20. Refer to notes for the re-prep procedure.
21. In cases where the software is unable to make a call: Select the Drop-down menu and change the analysis type from high confidence to high sensitivity. If the well color of the uncertain wells becomes either red or green, record the value in lab notebook.

22. Confirm the new call by looking at the second derivative curve (s).
23. Repeat steps with the FAM filter that will display the Target Cps.

3.3.3 Instrument Run and Analysis: Quant Studio 5

1. Tap touchscreen of instrument to wake.
2. Login to Cloud Account.
3. Enter Username:
4. Enter Password:
5. Load experiment template from Table 1.
6. Select QPCR test method.
7. In Experiment Properties tab, enter experiment file as: Example yyyymmdd_[location] QPCRID###_initials_qPCR.
8. Ensure that File is Saved to Cloud.
9. Ensure that both FAM (465–510 nm) and VIC/HEX (533–580 nm) excitation filters are selected.
10. Open tray.
11. Load 96-well plate.
12. Select start.
13. Once the run is complete, click the Transfer File tab on the instrument to save archive data.
14. On the computer, open and browser and login to Thermo-Fisher Cloud Account: thermofisher.com/mysso/loginDisplay.
15. Select App Display: https://apps.thermofisher.com/apps/spa/#/apps.
16. Select and Open the Design and Analysis Application.
17. Select Summary, then Multicomponent plot to view target results (Fig. 1).
18. Select Open Existing Experiment, then search for and open: .eds experiment file.
19. In the experiment manager, select Export tab, ensure that results are selected, then export results and download the excel file.
20. The file is downloaded as a zipped folder, unzip the folder, and locate the QPCR test results excel file.
21. In the excel home page, click on the Filter tab to enable filter function. Then, change the Target filter to Target 2 (VIC/HEX).

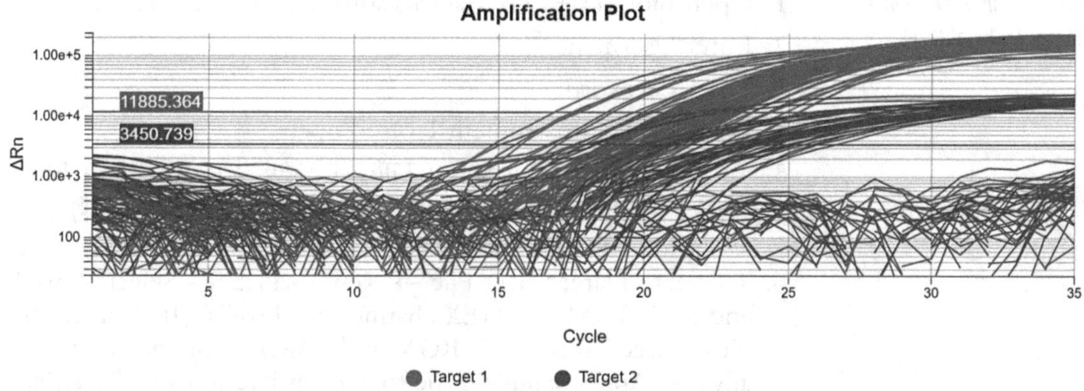

Fig. 1 Example amplification plot of the PACE assay. Fluorescent signals that cross the threshold are called positive for specific allele. Presence of only Target 1 (FAM) or Target 2 (HEX) corresponds to alternative homozygous alleles. Presence of FAM and HEX for a sample corresponds to heterozygous allele for the SNP target

22. Confirm the presence or absence of signal in the controls and then in the sample wells both by analyzing the Target 2 (VIC/HEX) Cp data.
23. View the amplification curves in the Results Tab in the Design and Analysis application looking at the shape for the cycle amplification curves for each well and the software demarcation.
24. Each fluorescent target threshold can be adjusted manually at your supervisor's discretion during analysis; however in general, the threshold line should be set to start to mid exponential phase of amplification curve.
25. Wells with detected signal above threshold line will be marked with Cp value for Target 1 (FAM) and Target 2 (VIC/HEX) and those with non-detectable signals will be undetermined.
26. NOTE: Samples with a positive signal will have a numerical value for Target 1 (FAM) once the Cp is calculated, whereas negative signal will be undetermined.
27. In cases of unexpected non-detects or undetermined for the Target 2 (VIC/Hex): Confirm your assessment of the Cp result by using the analysis features provided in the software.
28. If the curve is consistent with a non-detect result for both FAM and HEX, record the sample name in your lab notebook. The sample may need to be re-prepped to confirm the complete absence of the genotype.
29. Repeat steps with the Target 1 (FAM) filter that will display the Target 1 (FAM) Cps.
30. Save the Excel file with a unique filename and archive it.

3.3.4 Instrument Run and Analysis: Quant Studio 5

1. Open Biorad CFX96 Maestro software from the desktop.
2. Enter Username:
3. Enter Password:
4. Double Click Instrument Icon to open.
5. To Create Protocol: Select File → New→ Protocol → Enter PCR parameters from Table XX, ensuring plate read step at X annealing step. Save protocol file.
6. To create Plate: Select File → New → Plate → Select all wells and apply FAM and HEX channel to all wells. (If Virus Multiplex is used, Apply Cy5, ROX, and TAMRA channels additionally). → Apply Sample Type to appropriate wells → Save Plate file with unique qPCR Experiment barcode.
7. To setup Run: Select Run → Under Protocol Load appropriate file. → Under Plate Load appropriate file. → Under Start Run → Select Open Lid and Load plate on to instrument → Select Start Run.
8. Once the run is complete, save and archive datafile if not saved already.
9. In the Biorad CFX Maestro, open experiment datafile.
10. Confirm the presence or absence of signal in the controls and then in the sample wells both by analyzing the Target 2 (HEX) Cp data.
11. View the amplification curves in the Results Data Summary section looking at the shape/cycle of amplification curves for each well.
12. Each fluorescent target threshold can be adjusted manually at your supervisor's discretion during analysis; however in general, the threshold line should be set to start to mid exponential phase of amplification curve.
13. Wells with detected signal above threshold line will be marked with Cp value for Target 1 (FAM) and Target 2 (HEX) and those with non-detectable signals will be undetermined.
14. NOTE: Samples with a positive signal will have a numerical value for Target 1 (FAM) once the Cp is calculated, whereas negative signal will be N/A.
15. In cases of unexpected non-detects or undetermined for the Target 2 (Hex):
16. Confirm the assessment by looking at the amplification curve in the Results tab of the software.
17. If the curve is consistent with that of a non-detect, record the sample name in your lab notebook as the sample will need to be re-prepped.

18. Repeat **steps 3.3.4.10–3.3.4.11** with the Target 1 (FAM) filter that will display the Target 1 (FAM) Cps.
19. Save excel file with a unique filename and archive.

4 Notes

1. PACE oligonucleotides for target SNP must be well designed to produce a single amplicon product generally 100–150 bp and compatible with thermodynamic properties of nested oligonucleotide corresponding to FRET cassette. Oligonucleotide primer should target conserved regions surrounding a target genotype of interest or be uniquely designed for a given genotype target. The 3′ end of each forward primer must target a specific SNP for allele-specific detection. Two forward primers will be designed for amplification with a common reverse primer. The single nucleotide polymorphism detection will occur if the allele-specific primer is able to hybridize and produce an amplicon due to the presence of the target SNP. Target SNP corresponding primers must be nested with the following 5′ tails on the respective SNP targeting oligonucleotides: forward primer FRET FAM Fwd: GAA GGTGACCAAGTTCATGCT and FRET HEX Fwd: GAAGGTCGGAGTCAACGGATT. The FAM and HEX sequ ences will be used for each respective SNP.
2. A PACE Primer mix is prepared by mixing the following:
   10 μL PACE primer mix for PACE reactions
   (a) FAM oligonucleotide tail-allele-specific forward primer 100 μM Forward: 1.2 μL.
   (b) HEX oligonucleotide tail-allele-specific forward primer-FAM 100 μM Forward: 1.2 μL.
   (c) Common reverse oligonucleotide: 3 μL.
   (d) Nuclease-free water: 4.6 μL.
3. A 2 mm hole punch from leaf material or FTA printed samples can be used to produce precipitated nucleic acids for template input into PACE PCR. The size of the hole punch may be adjusted for higher DNA yields.
4. Ensure that all residual ethanol is removed after the wash step during PCR product cleanup. The amplicon product must be free of solvent contamination.
5. Each Pace Primer will detect SNPs at a single position. When preparing the PACE master mix, 0.138 μL of this primer mix will be prepared with 5 μL PACE2.0 Genotyping master mix per reaction for each PACE PCR reaction.

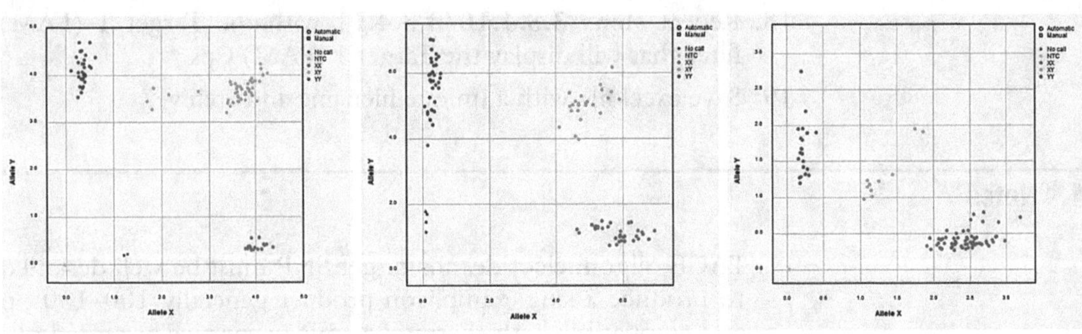

Fig. 2 Example scatter plots for the allele determination at three loci. *X* axis corresponds to fluorescence from samples homozygous with allele X (red data points) and *Y* axis corresponds to fluorescence from samples homozygous with allele Y (blue data points) and samples heterozygous with both X and Y alleles (green data points) for each SNP position

6. The 10-cycle stepwise decreasing ramp at 0.8 °C per cycle to a touchdown temperature of 57 °C is essential. Please design allele-specific oligonucleotides to accommodate annealing temperatures under these conditions.
7. For the FAM and HEX FRET cassette detection, ensure to utilize FAM 485/520 and HEX 520/560 filters to determine SNP profile for each assay built to determine target loci zygosity (Fig. 2).

Conflict of Interest

The authors have no conflicts of interest to declare that are relevant to the content of this chapter.

References

1. Abaya A, Zaro GC, De la Mora Pena A, Hsiang T, Goodwin PH (2024) Phenotypic and genotypic variation of cultivated Panax quinquefolius. Plants (Basel) 13(2). https://doi.org/10.3390/plants13020300
2. Anupama A, Bhugra S, Lall B, Chaudhury S, Chugh A (2018) Assessing the correlation of genotypic and phenotypic responses of indica rice varieties under drought stress. Plant Physiol Biochem 127:343–354. https://doi.org/10.1016/j.plaphy.2018.04.001
3. de Ronne M, Lapierre E, Torkamaneh D (2024) Genetic insights into agronomic and morphological traits of drug-type cannabis revealed by genome-wide association studies. Sci Rep 14(1):9162. https://doi.org/10.1038/s41598-024-58931-w
4. Kalendar R, Shustov AV, Akhmetollayev I, Kairov U (2022) Designing allele-specific competitive-extension PCR-based assays for high-throughput genotyping and gene characterization. Front Mol Biosci 9:773956. https://doi.org/10.3389/fmolb.2022.773956
5. Jang YJ, Oh Y, Verma S, Porter ME, Dalid C, Yoo C-M, Fan Z, Willborn CW, Han K, Kim D-S, Whitaker VM, Lee S (2024) Updates on strawberry DNA testing and marker-assisted breeding at the University of Florida. International Journal of Fruit Science 24(1):219–228. https://doi.org/10.1080/15538362.2024.2365683
6. Seeb JE, Carvalho G, Hauser L, Naish K, Roberts S, Seeb LW (2011) Single-nucleotide polymorphism (SNP) discovery and applications of SNP genotyping in nonmodel

organisms. Mol Ecol Resour 11(Suppl 1):1–8. https://doi.org/10.1111/j.1755-0998.2010.02979.x

7. Hayden MJ, Tabone T, Mather DE (2009) Development and assessment of simple PCR markers for SNP genotyping in barley. Theor Appl Genet 119(5):939–951. https://doi.org/10.1007/s00122-009-1101-7
8. Semagn K, Babu R, Hearne S, Olsen M (2013) Single nucleotide polymorphism genotyping using Kompetitive Allele Specific PCR (KASP): overview of the technology and its application in crop improvement. Mol Breed 33(1):1–14. https://doi.org/10.1007/s11032-013-9917-x
9. Kawash J, Erndwein L, Johnson-Cicalese J, Knowles S, Vorsa N, Polashock J (2024) Quantitative trait loci analysis and marker development for fruit rot resistance in cranberry shows potential genetic association with epicuticular wax. Phytopathology 114(6):1366–1372. https://doi.org/10.1094/PHYTO-12-23-0477-R
10. Baloch FS, Altaf MT, Liaqat W, Bedir M, Nadeem MA, Comertpay G, Coban N, Habyarimana E, Barutcular C, Cerit I, Ludidi N, Karakoy T, Aasim M, Chung YS, Nawaz MA, Hatipoglu R, Kokten K, Sun HJ (2023) Recent advancements in the breeding of sorghum crop: current status and future strategies for marker-assisted breeding. Front Genet 14:1150616. https://doi.org/10.3389/fgene.2023.1150616
11. Di Nunzio M, Barrot-Feixat C, Gangitano D (2024) Characterization and evaluation of nine Cannabis sativa chloroplast SNP markers for crop type determination and biogeographical origin on European samples. Forensic Sci Int Genet 68:102971. https://doi.org/10.1016/j.fsigen.2023.102971
12. Dowling CA, Shi J, Toth JA, Quade MA, Smart LB, McCabe PF, Schilling S, Melzer R (2024) A FLOWERING LOCUS T ortholog is associated with photoperiod-insensitive flowering in hemp (Cannabis sativa L.). Plant J 119(1):383–403. https://doi.org/10.1111/tpj.16769
13. Toth JA, Stack GM, Carlson CH, Smart LB (2022) Identification and mapping of major-effect flowering time loci Autoflower1 and Early1 in Cannabis sativa L. Front Plant Sci 13:991680. https://doi.org/10.3389/fpls.2022.991680
14. Leckie KM, Sawler J, Kapos P, MacKenzie JO, Giles I, Baynes K, Lo J, Baute GJ, Celedon JM (2024) Loss of daylength sensitivity by splice site mutation in Cannabis pseudo-response regulator. Plant J 118(6):2020–2036. https://doi.org/10.1111/tpj.16726

Chapter 13

Power of Short Tandem Repeat (STR) Analysis for Analysis of Genetic Differences and Genotype Identification

Anthony Torres and Reginald Gaudino

Abstract

This chapter presents a protocol for short tandem repeat (STR) analysis, a robust, economical, and highly accurate method commonly used in forensics and agriculture for genetic profiling and identity determination. The protocol involves DNA isolation from FTA cards or leaf pieces, followed by PCR amplification of STR regions and fragment analysis. The number of repeats in these short tandem patterns is unique to individuals, making it useful for genotyping analysis. By analyzing STR profiles from reference samples, a reference STR table is generated, serving as a comparison tool for identifying repeat differences between the genotypes of test individuals. This comprehensive analysis of genetic variations provides valuable insights for differentiating between individuals based on their unique genetic profiles.

Key words Short tandem repeat (STR) analysis, Genetic profiling, Genotyping, PCR, Amplicon fragment analysis, FTA cards, Leaf tissue, DNA, Amplicon, Fragment, STR, Repeats, Primers, Lineage

1 Introduction

Short tandem repeat (STR) analysis has emerged as a powerful and cost-effective tool in genetic profiling, finding widespread applications in diverse fields such as forensic science, agriculture, and conservation biology [1–4]. This technique exploits the specific inherited variability in the number of short tandem repeated sequences within DNA, providing a unique genetic fingerprint for individual identification, parentage, and differentiation [5]. STR analysis typically involves PCR amplification of target STR loci from DNA templates, followed by fragment analysis using capillary electrophoresis [6, 7]. The high discriminatory power and reproducibility of STR profiling have established it as a cornerstone for various applications, including cultivar identification of *C. sativa* [8, 9]. Notably, STR analysis has played a key role in cultivar identification of *Cannabis sativa*, enabling accurate differentiation between strains and aiding in regulatory compliance [10].

Anthony Torres (ed.), *Genotyping: Methods and Protocols*, Methods in Molecular Biology, vol. 2943,
https://doi.org/10.1007/978-1-0716-4642-7_13,

Recent technological advancements, including high-throughput platforms like the SeqStudio Genetic Analyzer, have further streamlined the STR analysis workflows in molecular biology laboratories, making it more accurate and efficient to carry out short tandem repeat genotyping and analysis to build reference databases, crucial for comparative analysis and identification purposes [1, 11]. Ongoing method development for genotyping to analyze genetic variation shows that STR genotyping is a crucial tool for both fundamental and applied research. A standard workflow was developed for isolating DNA from FTA cards, followed by PCR and size fragment analysis genotyping of *C. sativa* using established SSR markers [12, 13].

2 Materials

2.1 Sample Preparation

1. Bunsen burner.
2. Calibrated p1, p100, p200 multichannel pipettes and p1, p200, p1000 single-channel pipettes.
3. 100–1000 μL nuclease-free aerosol filtered pipette tips.
4. 20–300 μL nuclease-free aerosol filtered pipette tips.
5. 1–100 μL nuclease-free aerosol filtered pipette tips.
6. 0.1–20 μL nuclease-free aerosol filtered pipette tips.
7. Stainless steel dissecting forceps.
8. 250 mL beaker.
9. Optional: plate piercer.
10. 1.5 mL centrifuge tubes.
11. Gloves.
12. 100% ethanol.
13. Nuclease-free water.
14. Extraction buffer solution: 0.25 M Tris–HCl, pH 8, 0.25% Triton-X.
15. Salt buffer solution: 5 M NaCl.
16. FTA plant saver cards.
17. Unicore punch kit, 2 mm.

2.2 Fluorescently Labeled Primer-STR-PCR

1. Standard thermocycler.
2. Mini plate spinner.
3. Gloves.
4. Clean designated space.
5. Clear 96-well plate or similar.
6. Sterile loading trough.

7. 5′ Fluorophore-labeled oligonucleotides for STR analysis (*see* **Note 1**).
8. GoTaq PCR Mix (Promega).
9. Genotype-specific positive DNA for the positive control template.
10. AlumaSeal II sealing foils.

2.3 Genetic Analyzer Preparation of PCR Product

1. Hi-Di formamide (Thermo Fisher) (*see* **Note 2**).
2. 96-well plate.
3. GeneScan™ 400HD ROX™ dye Size Standard (Thermo Fisher) (*see* **Note 3**).

2.4 Loading Genetic Analyzer and Size Fragment Analysis of STR PCR Products

1. SeqStudio™ Cartridge (Thermo Fisher).
2. SeqStudio Genetic Analyzer Cathode Buffer Container (Thermo Fisher).
3. 96-well Septa for SeqStudio™ Genetic Analyzer (Thermo Fisher).

3 Methods

3.1 Sample Preparation

1. Prepare the workstation on the lab bench by wiping down the surface with alcohol where the preparation will be performed.
2. Prepare a clean PCR setup with clean 1.5 mL microcentrifuge tubes for sample preparation in an easily accessible area.
3. Sterilize forceps by dipping them in alcohol and flame sterilizing them or aseptically wiping excess alcohol until all alcohol has evaporated.
4. Remove a small sampling of raw tissue pieces or imprinted FTA card (~4–8 mm) with flame-sterilized forceps or alcohol-sterilized QIAGEN 2 mm UniCore Punch tool, taking care not to touch the sample directly with gloves. (*see* **Note 4**).
5. Carefully place the sample into a 1.5 mL microcentrifuge tube at the bottom of the well so that it does not contact the top of the tube.
6. Repeat **steps 3–5** with remaining lots/replicates until all samples are processed.
7. Fill the sterile liquid basin/trough with Extraction Buffer using a dedicated single-channel or multichannel 30–300 μL multichannel pipettor loaded with clean tips to draw and dispense 200 μL of extraction buffer into each tube.
8. Repeat for all remaining samples and replace tips each time—cap tubes carefully in between steps.

9. Spin tubes down in microcentrifuge for 60 s to ensure all material is fully immersed in the extraction buffer.
10. Load the tube on a thermal incubation block and set to run 10 min "Boil" with lids sealed with tube lock to close.
11. ~95–100 °C 10 min 1× (*see* **Note 5**).
12. Following Boil, add 8 μL 5 M NaCl and vortex.
13. Add 400 μL 100% ethanol and invert several times.
14. Place tubes into −80 °C for 15 min or −20 °C overnight to precipitate DNA.
15. Following ultra-cold precipitation, mark the position where pellet will form onto tube and centrifuge 1.5 mL tubes for 30 min at 4 °C.
16. Remove tubes carefully, decant the supernatant, and retain the pellet.
17. Optional: Perform a 70% ethanol wash to remove additional residual waste.
18. Following wash, centrifuge 1.5 mL tubes at 12,000 × *g* or greater for 30–60 min at 4 °C.
19. Remove tubes carefully, decant the supernatant, and retain the pellet (*see* **Note 6**).
20. Allow pellets to dry for 5–10 min at room temperature, then reconstitute the dried pellets in 50 μL nuclease-free water.

3.2 Fluorescently Labeled Primer-STR-PCR Setup

1. Prepare clean STR PCR setup in an easily accessible area.
2. Prepare five duplex STR reactions with 25 μL final reactions with GoTaq Master Mix and Custom labeled STR Primer Mix according to the following:

Reagent	Volume for 1 reaction (μL)
GoTaq Master Mix	12.5
5′ Labeled Primer Mix (10 μM)	2.4 (*see* **Note 7**)
Nuclease-free water	9.1

3. In the dedicated PCR space, use a 0.1–10 μL multichannel pipette, load 0.1–20 μL filtered pipette tips, pipette 24 μL of STR Master Mix PCR reaction mix to each well, respectively, for each sample (*see* **Note 6**).
4. Change tips and prepare to load samples, use fresh tips for each new sample.
5. For all reactions tested with reconstituted DNA input, using a 0.1–10 μL multichannel pipette or single-channel pipette, draw and dispense 1 μL of reconstituted DNA input. Be sure tips are

below the surface of the mix before dispensing. Mix by pipetting up and down several times, then eject tips after the final mix.

6. For all reactions with purified genomic DNA, using a 0.1–10 μL multichannel pipette or single-channel pipette, draw and dispense 1 μL of kit (1–10 ng/μL)* purified genomic DNA from DNEasy or similar, ensure that tips are below surface of mix before dispensing, mix by pipetting up and down several times, then, eject tips after the final mix.
7. After each sample well has been filled, obtain the control standards and aliquot 1 μL of positive control template, negative control template, and no template control into their own appropriate wells.
8. Seal plate thoroughly with AlumaSeal sealing foil.
9. Spin plates for 10 s.
10. The plate is now ready for PCR.
11. Place the PCR plate in the real-time instrument and select and run the.
12. Fluorescently labeled primer-STR-PCR protocol:
    (a) 95 °C for 5 min 1×.
    (b) 95 °C for 15 s.
    (c) 55 °C for 30 s (*see* **Note 8**).
    (d) 72 °C for 30 s.
    (e) Repeat **steps a–c** 40×.
13. After STR PCR reaction, pull plate out of thermocycler and allow to acclimate to room temperature, then prepare a 1:10 working dilution of the PCR product.
14. In a fresh ABI PCR plate, prepare 9.5 μL Hi-Di formamide + 1 μL of the 1:10 STR PCR product + 0.5 μL of the GeneScan™ 400HD ROX™ dye Size Standard, mix well by pipetting slowly.
15. Prepare the plate so that four reactions/duplexes will be injected at one time (as a single SeqStudio cartridge injection is four reactions.)
16. One the wells are prepared, seal the plate with a 96-well Septa for SeqStudio™ Genetic Analyzer, and load the plate into the Seq Studio for Genotype Fragment Analysis.

3.3 Loading Genetic Analyzer and Size Fragment Analysis of STR PCR Products

1. Tap touchscreen of SeqStudio Instrument to wake.
2. Login to Cloud Account.
3. Enter Username/Password.

4. Ensure SeqStudio Genetic Analyzer Cartridge has enough injections and is not expired.
5. Ensure SeqStudio Genetic Analyzer Cathode Buffer Container is fresh and is not expired.
6. Open tray.
7. Load 96-well plate.
8. Select start.
9. Select setup new experiment.
10. Select fragment analysis.
11. Select long fragment analysis and chose wells that will be injected into the Seq Studio cartridge.
12. Select run plate.
13. When run is completed, transfer .fsa files to Thermo Fisher Cloud.
14. Open MSA cloud software.
15. Start New Project and Load in .fsa files from genotyping experiment.
16. Ensure that file is saved to Cloud.
17. In the MSA Thermo Fisher cloud software Setup tab, select Size Standard (sizes from GeneScan™ 400HD ROX™ dye Size Standard), load Allele Panel if applicable to perform Genotyping Analysis, and select analyze samples (*see* **Note 9**).
18. In the Results tab, a quality analysis will be performed to give the called size an SQ score and green, yellow, or red status.
19. Toggle between sizes to select peaks appropriate for each well. To account for each of the short tandem repeat elements from the duplex, a low-size and high-size set of sizes should be visualized.
20. To visualize genotyping results, analyze sizing, genotyping, and size matching to visualize fragment size for each sample/amplicon set (Fig. 1).
21. Refer to references and analyze short tandem repeat size of the amplified size-specific allele for each genotype and compare against reference sample and size standard (Fig. 2).

4 Notes

1. Primers must be well-designed and specific for PCR amplification of DNA regions containing the size-specific tandem repeat element. The size of products must be considered and well separated or distinguished with unique 5′ fluorophore in

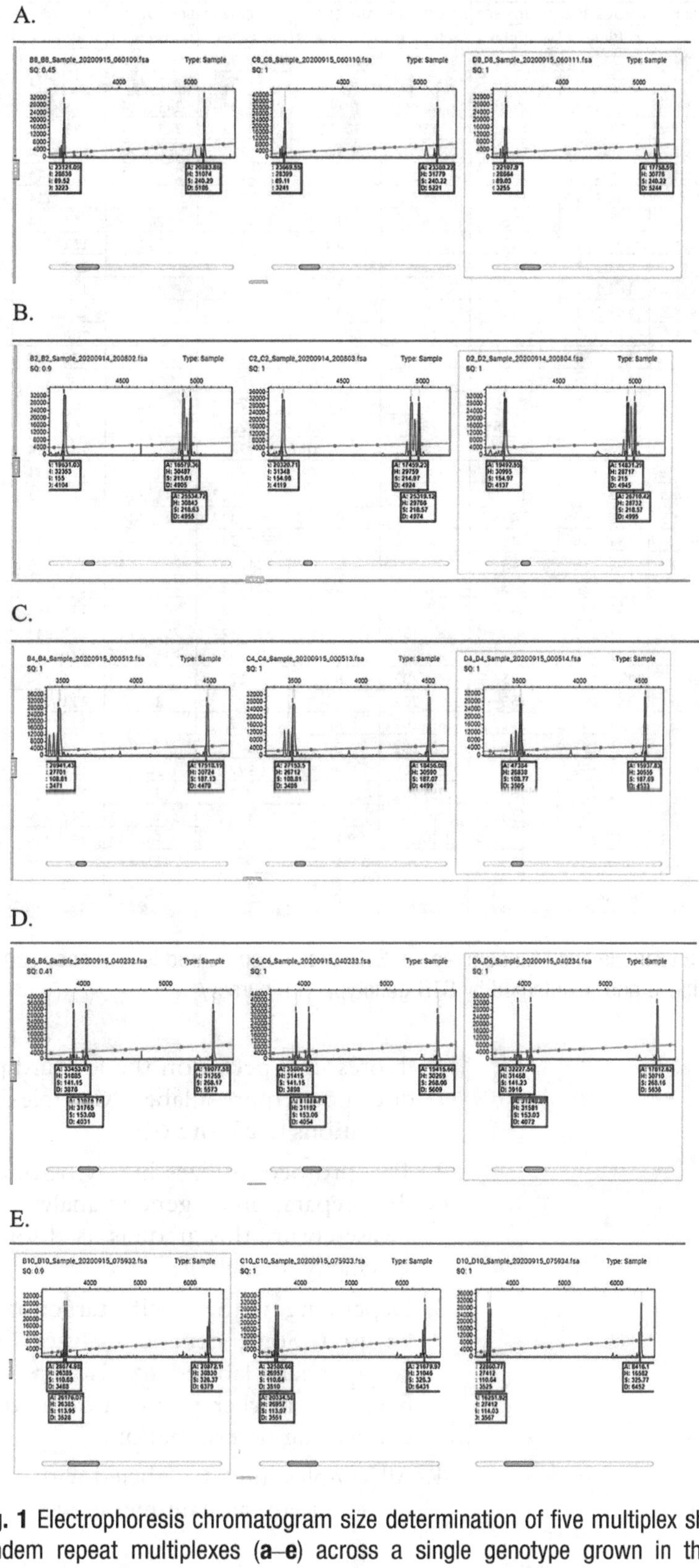

Fig. 1 Electrophoresis chromatogram size determination of five multiplex short tandem repeat multiplexes (**a**–**e**) across a single genotype grown in three locations and confirmation of genetic identify

Sample ID	ANUCS 302	ANUCS 305	ANUCS 308	ANUCS 501	B01-CANN1	B05-CANN1	D02-CANN1	E07-CANN1	H06-CANN2	H-09 CANN 2
	Multiplex 4	Multiplex 2	Multiplex 3	Multiplex 1	Multiplex 5	Multiplex 1	Multiplex 5	Multiplex 3	Multiplex 4	Multiplex 2
PUC Ref 1	141	155	185, 202	90, 100	326, 327	241	107, 114	106, 112	268, 271	215, 219
MM Ref	141, 153	155	200, 202	90, 95	326, 327	231, 244	111, 114	106	274	205, 219
BS Test Ref	141, 143	155	185	90, 95	328, 340	241, 244	113, 114	106, 112	271, 274	215, 225
RH Test Ref	141, 153	155	185	90, 100	326, 327	241, 244	113, 114	106, 112	271	215, 221
PUC Test 2	141	155	185, 202	90	327	241, 244	113, 114	106	271, 274	215, 219
PUC Test 4	141	155	185, 202	90	327	241, 244	113, 114	106	271, 274	215, 219
GSC Ref 1	141,153	155	185	90	327	241	111, 114	106	271	205, 215
GSC Test 1	141,153	155	185	90	327	241	111, 114	106	271	205, 215
GSC Test 2	141,153	155	185	90	327	241	111, 114	106	271	205, 215
GSC Test 3	141,153	155	185	90	327	241	111, 114	106	271	205, 215
GSC Test 4	141,153	155	185	90	327	241	111, 114	106	271	205, 215
GSC Test 5	141,153	155	185	90	327	241	111, 114	106	271	205, 215
BBM Ref 1	147, 153	149, 155	185	90, 100	326	241	108, 114	109	268, 274	217, 219
BBM Test 1	147, 153	149, 155	185	90, 100	326	241	108, 114	109	268, 274	217, 219
BBM Test 2	147, 153	149, 155	185	90, 100	326	241	108, 114	109	268, 274	217, 219
BBM Test 3	147, 153	149, 155	185	90, 100	326	241	108, 114	109	268, 274	217, 219
BBM Test 4	147, 153	149, 155	185	90, 100	326	241	108, 114	109	268, 274	217, 219
BBM Test 5	147, 153	149, 155	185	90, 100	326	241	108, 114	109	268, 274	217, 219
PUC Test 1	141, 156	143	185, 200	90	324, 327	241, 244	110, 111	106	271	215, 219
PUC Test 3	141, 156	143	185, 200	90	324, 327	241, 244	110, 111	106	271	215, 219
PUC Test 5	141, 156	143	185, 200	90	324, 327	241, 244	110, 111	106	271	215, 219
GE Ref	141, 156	143	185, 200	90	324, 327	241, 244	110, 111	106	271	215, 219
GE Test 1	141, 156	143	185, 200	90	324, 327	241, 244	110, 111	106	271	215, 219
GE Test 2	141, 156	143	185, 200	90	324, 327	241, 244	110, 111	106	271	215, 219
GE Test 3	141, 156	143	185, 200	90	324, 327	241, 244	110, 111	106	271	215, 219
GE Test 4	141, 156	143	185, 200	90	324, 327	241, 244	110, 111	106	271	215, 219
GE Test 5	141, 156	143	185, 200	90	324, 327	241, 244	110, 111	106	271	215, 219
CG Ref	141, 153	155	185, 202	90	327	241	111, 114	103, 106	274	205, 215
CG Test 1	141, 153	155	185, 202	90	327	241	111, 114	103, 106	274	205, 215
CG Test 2	141, 153	155	185, 202	90	327	241	111, 114	103, 106	274	205, 215
CG Test 3	141, 153	155	185, 202	90	327	241	111, 114	103, 106	274	205, 215
CG Test 4	141, 153	155	185, 202	90	327	241	111, 114	103, 106	274	205, 215
CG Test 5	141, 153	155	185, 202	90	327	241	111, 114	103, 106	274	205, 215
MM Test 1	141, 153	155	200, 202	90, 95	324	241, 244	111, 114	106, 112	271	205, 219
MM Test 2	141, 153	155	200, 202	90, 95	324	241, 244	111, 114	106, 112	271	205, 219
MM Test 3	141, 153	155	200, 202	90, 95	324	241, 244	111, 114	106, 112	271	205, 219
MM Test 4	141, 153	155	200, 202	90, 95	324	241, 244	111, 114	106, 112	271	205, 219
MM Test 5	141, 153	155	200, 202	90, 95	324	241, 244	111, 114	106, 112	271	205, 219

Fig. 2 Example short tandem repeat (STR) profiles compared to reference sample for multiple plant genotypes cataloged and maintained by STR genotype in a nursery

fluorescent spectra on the forward primer. The reverse primer does not require a label. Multiplexing is possible under these conditions (*see* **Note 6**).

2. This protocol utilizes highly deionized (Hi-Di) formamide in the preparation of genetic analyzer reaction of PCR product. Please ensure this grade is used for the best resolution of size specificity in analysis.
3. Depending on the specific target size, different size ladders can be used. Some genetic analyzers have preset parameters for commercially labeled size ladders. The standard curve should have size marker points around the target size for accurate genotyping determination.
4. All samples must be labeled with a unique identification number, and an alphanumeric well position number must be

assigned for all future labeling. Additionally, if FTA imprinted spots are light, use greater sampling size (~8 mm) in same extraction volume (100 μL).

5. Subheading 3.1, **steps 5–11** can be carried out in a 96-well plate for higher sample throughput. However, to complete DNA purification using sodium ethanol precipitation, high-speed centrifugation is required (>12,000 × *g*), and samples can be transferred to 1.5 microcentrifuge tubes at this point.
6. It is important to ensure that no residual ethanol remains before reconstitution, as it can inhibit PCR downstream. Make sure enough time has been allowed to dry the pellet. As you marked the expected position of the pellet, you will be able to track it through the steps. However, sometimes it can be invisible to the naked eye, so be sure to wash all sides of the tube when you reconstitute the DNA.
7. Custom fluorophore-labeled primer mixes can be prepared individually from your designed specific targets, duplex, or multiplex PCR. As with oligo design, when preparing a primer mixture, ensure that the fluorophore-labeled primer sets are compatible with each other and do not induce cross-reactivity. Also, for size analysis, keep in mind that amplicon products must be resolved or labeled properly to allow for efficient resolution.
8. Optimized oligonucleotide concentrations and compatible annealing temperatures will help PCR efficiency, especially in multiplex.
9. When performing sizing analysis on multiple samples, it is important to use standard size and genotype references and saved size calling parameters to ensure accuracy and reproducibility of size determinations for your target genotypes.

Conflict of Interest

The authors have no conflicts of interest to declare that are relevant to the content of this chapter.

References

1. Qu W, Wang F, Sun D, Liu Y, Jin X, Gong Z, Liu J, Zhang N, Wen D, Wang C, Jia H, Xu R, Tang X, Chen S, Fu X, Li X, Rong H, Zhang T, Jin C, Chen L, Li J, Liu Y, Cai J, Jiang B, Zha L (2024) Internal validation of the GA118-24B Genetic Analyzer, a stable capillary electrophoresis system for forensic DNA identification. Int J Legal Med 138(2):361–373. https://doi.org/10.1007/s00414-023-03106-x
2. Vasek J, Cilova D, Melounova M, Svoboda P, Zdenkova K, Cermakova E, Ovesna J (2021) OpiumPlex is a novel microsatellite system for profiling opium poppy (Papaver somniferum L.). Sci Rep 11(1):12799. https://doi.org/10.1038/s41598-021-91962-1
3. Cateau E, Leclerc A, Cartier N, Valsecchi I, Bailly E, Le Senechal R, Becerra M, Le Gallou B, Lavergne RA, Chesnay A, Robin JP,

Cray C, Goddard N, Thorel M, Guillot J, Mulot B, Desoubeaux G (2022) Aspergillosis in a colony of Humboldt penguins (Spheniscus humboldti) under managed care: a clinical and environmental investigation in a French zoological park. Med Mycol 60(7):myac046. https://doi.org/10.1093/mmy/myac046

4. Testolin R, Messina R, Cipriani G, De Mori G (2023) SSR-based DNA fingerprinting of fruit crops. Crop Sci 63(2):390–459. https://doi.org/10.1002/csc2.20896
5. Driskill M, Pardee K, Hummer KE, Zurn JD, Amundsen K, Wiles A, Wiedow C, Patzak J, Henning JA, Bassil NV (2022) Two fingerprinting sets for Humulus lupulus based on KASP and microsatellite markers. PLoS One 17(4):e0257746. https://doi.org/10.1371/journal.pone.0257746
6. Wang J, Wang W, Liu Y, Duo L, Huang L, Jiang X (2009) The method of single-nucleotide variations detection using capillary electrophoresis and molecular beacons. Mol Biol Rep 36(7):1903–1908. https://doi.org/10.1007/s11033-008-9397-y
7. Mansfield ES, Vainer M, Enad S, Barker DL, Harris D, Rappaport E, Fortina P (1996) Sensitivity, reproducibility, and accuracy in short tandem repeat genotyping using capillary array electrophoresis. Genome Res 6(9):893–903. https://doi.org/10.1101/gr.6.9.893
8. Fisarova L, Surinova M, Jarosova A, Krejcik J, Vosatka M (2024) Evidence of the ability of microsatellite method to distinguish Cannabis strains with high cannabinoid content. Cannabis Cannabinoid Res 9(2):513–522. https://doi.org/10.1089/can.2022.0272
9. Di Nunzio M, Barrot-Feixat C, Gangitano D (2024) Characterization and evaluation of nine Cannabis sativa chloroplast SNP markers for crop type determination and biogeographical origin on European samples. Forensic Sci Int Genet 68:102971. https://doi.org/10.1016/j.fsigen.2023.102971
10. Howard C, Gilmore S, Robertson J, Peakall R (2009) A Cannabis sativa STR genotype database for Australian seizures: forensic applications and limitations. J Forensic Sci 54(3):556–563. https://doi.org/10.1111/j.1556-4029.2009.01014.x
11. Nishio S, Kunihisa M, Taniguchi F, Kajiya-Kanegae H, Moriya S, Takeuchi Y, Sawamura Y (2021) Development of SSR databases available for both NGS and capillary electrophoresis in apple, pear and tea. Plants (Basel) 10(12):2796. https://doi.org/10.3390/plants10122796
12. Kohnemann S, Nedele J, Schwotzer D, Morzfeld J, Pfeiffer H (2012) The validation of a 15 STR multiplex PCR for Cannabis species. Int J Legal Med 126(4):601–606. https://doi.org/10.1007/s00414-012-0706-6
13. Xia R, Tao R, Qu Y, Zhang X, Yu H, Yuan C, Zhang S, Li C (2022) Development and validation of a novel and fast detection method for Cannabis sativa: a 19-Plex short tandem repeat typing system. Front Plant Sci 13:837945. https://doi.org/10.3389/fpls.2022.837945

Chapter 14

Targeted Amplicon Genotyping by Sanger Sequencing

Anthony Torres and Reginald Gaudino

Abstract

This chapter presents a protocol for Sanger sequencing, a conventional yet rapid, cost-effective, and highly accurate method for targeted sequencing. It involves designing specific primers for forward and reverse template extension reactions, during which fluorophore labeled, chain terminating dideoxynucleotides are incorporated into newly replicated DNA molecules. The resulting end-labeled terminated products are separated by electrophoresis, enabling precise base identification within the amplicon. Chromatogram analysis further allows for the detection of nucleotide changes that may correspond to amino acid variations in targeted genes from different samples.

Key words DNA, Amplicon, Sanger sequencing, End termination reaction, Single nucleotide polymorphism (SNP), Targeted sequencing, SNP detection, Electrophoresis, Chromatogram analysis

1 Introduction

Sanger sequencing for genotyping is a flexible, cost-effective, and rapid approach for identifying single nucleotide changes in target amplicons [1, 2]. It is conventional and widely used as it is accurate and adequate for genotyping a wide range of applications [3–5]. There have been many adaptations using this approach; however, in principle, the method utilizes polymerase chain reaction extension of a given amplicon product by incorporating a fluorescently labeled nucleotide and terminating the reaction of an extending amplicon product [6]. As the products of different sizes accumulate, each with an end-terminal fluorophore, this collection of amplicons can be separated by microcapillary electrophoresis, and the fluorophore signal can be measured [7].

Well-designed and specific primers are necessary to target a given DNA or amplicon region, and it is also essential to incorporate an M13 oligonucleotide addition in the design of the forward and reverse primers for fluorophore incorporation and terminated extension [8]. In this study, we added M13 oligonucleotides to target primer sets from three different genes to rapidly carry out an

Anthony Torres (ed.), *Genotyping: Methods and Protocols*, Methods in Molecular Biology, vol. 2943,
https://doi.org/10.1007/978-1-0716-4642-7_14,

efficient sanger sequencing genotyping workflow end-to-end from sample input, through target PCR enrichment using M13 tagged primers, and cycle sequencing fluorophore incorporation and terminated extension, sequencing reaction clean up, and preparation of reaction loading for capillary analysis using a Seq Studio standard set of reagents [9, 10]. Genotyping results are resolved in chromatogram analysis [11]. In this workflow, plant samples were genotyped for selection using a routine targeted Sanger sequencing workflow to confirm single nucleotide base changes in three different genes.

2 Materials

2.1 Sample Preparation

1. Bunsen burner.
2. Calibrated p1, p100, p200 multichannel pipettes and p1, p200, p1000 single-channel pipettes.
3. 100–1000 μL nuclease-free aerosol filtered pipette tips.
4. 20–300 μL nuclease-free aerosol filtered pipette tips.
5. 1–100 μL nuclease-free aerosol filtered pipette tips.
6. 0.1–20 μL nuclease-free aerosol filtered pipette tips.
7. Stainless steel dissecting forceps.
8. 250 mL beaker.
9. Optional: plate piercer.
10. 1.5 mL centrifuge tubes.
11. Gloves.
12. Isopropanol.
13. Nuclease-free water.
14. Extraction buffer solution: 0.25 M Tris–HCl, pH 8, 0.25% Triton-X.
15. Salt buffer solution: 5 M NaCl.
16. FTA plant saver cards.
17. Unicore punch kit, 2 mm.

2.2 Big Dye Direct and Cycle Sequencing PCR

1. Standard thermocycler.
2. Genotype-specific positive DNA for positive control template.
3. Plastic plate sealer.
4. Mini plate spinner.
5. Clear 96-well plates or similar.
6. AlumaSeal II sealing foils.
7. Big Dye Direct Master Mix (Thermo Fisher).

8. BigDye Direct (BDD) Cycle Sequencing kit (Thermo Fisher, Fremont, CA, USA).
9. Custom M13 tailed oligonucleotides for Sanger sequencing analysis.

2.3 Sequencing Assays on the SeqStudio

1. SeqStudio Capillary Genetic Analyzer (Thermo Fisher).
2. Sterile loading trough.
3. Digital Vortex-Genie® 2.
4. SeqStudio genetic analyzer cathode buffer container (Thermo Fisher).
5. SeqStudio™ Cartridge (Thermo Fisher).
6. Big Dye Direct Terminator X (Thermo Fisher).
7. 96-well Septa for SeqStudio™ genetic analyzer (Thermo Fisher).

3 Methods

3.1 Sample Preparation

1. Prepare the workstation on the lab bench by wiping the surface with alcohol where the preparation will be performed (*see* **Note 1**).
2. Prepare a clean PCR setup with clean 1.5 mL microcentrifuge tubes in an easily accessible area for sample preparation.
3. Sterilize forceps by dipping them in alcohol and flame sterilizing them or aseptically wiping excess alcohol until all alcohol has evaporated.
4. Remove a small sampling of raw tissue pieces or imprinted FTA card (~4–8 mm) with flame-sterilized forceps or alcohol-sterilized QIAGEN 2 mm UniCore punch tool, taking care not to touch the sample directly with gloves (*see* **Note 2**).
5. Carefully place the sample into a 1.5 mL microcentrifuge tube at the bottom of the well so that it does not contact the top of the tube.
6. Repeat **steps 3–5** with remaining lots/replicates until all samples are processed.
7. Fill the sterile liquid basin/trough with Extraction Buffer using a dedicated single-channel or multichannel 30–300 μL multichannel pipettor loaded with clean tips to draw and dispense 200 μL of extraction buffer into each tube.
8. Repeat for all remaining samples, replacing tips each time and ensuring tubes are capped between steps.
9. Spin tubes down in microcentrifuge for 60 s to fully immerse all material in the extraction buffer.

10. Load the tube on a thermal incubation block and set to run for 10 min. Boil with lids sealed with tube lock to close.
    (a) ~95–100 °C 10 min 1×.
11. Following boil, add 8 μL 5 M NaCl and vortex.
12. Add 400 μL 100% isopropanol and invert several times.
13. Place tubes at −80 °C overnight for 15 min or −20 °C to precipitate DNA.
14. Following ultra-cold precipitation, mark the position where the pellet will form onto the tube and centrifuge 1.5 mL tubes for 30 min at 4 °C.
15. Remove tubes carefully, decant the supernatant, and retain the pellet.
    (a) Optional: Perform a 70% ethanol wash to remove additional residual waste.
16. Following wash, centrifuge 1.5 mL tubes for 30 min at 4 °C.
17. Remove tubes carefully, decant the supernatant, and retain the pellet.
18. Allow pellets to dry for 5–10 min at room temperature, then reconstitute the dried pellets in 20 μL of nuclease-free water.

3.2 Big Dye Cycle Sequencing PCR Reaction

1. Prepare a clean PCR setup in an easily accessible area.
2. Prepare a Big Dye Direct PCR Reaction (*see* **Notes 3** and **4**) according to:
   Reagent:
   Big Dye Direct PCR Master Mix: 5 μL.
   M13 tailed primer mix (0.8 μM): 1.5 μL.
   Nuclease-free water: 2.5 μL.
3. In the dedicated PCR space, use a 0.1–10 μL multichannel pipette, load 0.1–20 μL filtered pipette tips, pipette 9 μL of Big Dye Direct Master Mix PCR reaction mix with M13 tailed primers to each well, respectively, for each sample (*see* Subheading 4).
4. Change tips and prepare to load samples, use fresh tips for each new sample.
5. For all reactions tested with reconstituted DNA input, using a 0.1–10 μL multichannel pipette or single-channel pipette, draw and dispense 1 μL of reconstituted DNA input ensure that tips are below surface of mix before dispensing, mix by pipetting up and down several times, then, eject tips after final mix. This brings the reaction volume to a Big Dye cycle sequencing PCR reaction to 10 μL.
6. For all reactions with purified genomic DNA, using a 0.1–10 μL multichannel pipette or single-channel pipette,

draw and dispense 1 µL of kit (4–10 ng/µL)* purified genomic DNA from DNEasy or similar, ensure that tips are below the surface of the reaction mix before dispensing, mix by pipetting up and down several times, then, eject tips after final mix (*see* **Note 5**).

7. After filling each sample well, obtain the control standards and aliquot 1 µL of positive genotype control template, negative control template, and no template control into their own appropriate wells.
8. Seal the plate thoroughly with AlumaSeal sealing foil.
9. Spin plates for 10 s.
10. The plate is now ready for PCR.
11. Place the cycle sequencing PCR plate in a standard thermos cycler instrument and select and run the Big Dye direct PCR protocol:
    (a) 95 °C for 10 min 1×.
    (b) 95 °C for 3 s.
    (c) 60 °C for 15 s (*see* **Note 6**).
    (d) 68 °C for 30 s.
    (e) Repeat **steps b–d** 35×.
    (f) 72 °C for 2 min 1×.
12. After Big Dye direct PCR reaction, prepare two cycle sequencing (chain-end termination) reactions: a forward and reverse cycle sequencing master mix accordingly (*see* **Note 7**):
    Reagent:
    Big Dye Direct Cycle Sequencing Master Mix: 2 µL.
    M13 tailed forward primer (0.8 µM): 1 µL.
    Or
    M13 tailed reverse primer (0.8 µM): 1 µL.
13. Pull the plate out of the thermocycler and allow it to acclimate to room temperature. Then, prepare the PCR product and add 3 µL of either forward or reverse cycle sequencing master mix to the appropriate sample.
14. Seal plate thoroughly with AlumaSeal sealing foil and spin plates for 10 s. The plate is now ready for PCR.
15. Place the cycle sequencing PCR plate in the thermocycler instrument and run the Big Dye Direct Cycle Sequencing Protocol as detailed in the next step.
16. Big Dye Direct Cycle Sequencing protocol:
    (a) 37 °C for 15 min 1×.
    (b) 80 °C for 2 min 1×.
    (c) 96 °C for 1 min 1×.

(d) 96 °C for 10 s.
(e) 50 °C for 5 s.
(f) 60 °C for 75 s.
(g) Repeat **steps d–f** 25×.
(h) 4 °C hold.

17. Following cycle sequencing reaction, purify the reaction with the Big Dye Xterminator (Thermo Fisher) Clean Up Kit.
18. Prepare a working stock solution of SAM™ solution and XTerminator® solution with 45 μL SAM™ solution and 10 μL XTerminator® solution for 55 μL total per reaction. Scale up according to the number of reactions, ensure to vortex XTerminator® Solution immediately before pipetting (*see* **Note 8**).
19. Keep pipetting and mixing working stock before use.
20. Add 55 μL of working stock solution to each well and seal with AlumaSeal.
21. Plate plate on Digital Vortex-Genie® 2 and mix at 1800 rpm for 20 min (*see* **Note 9**).
22. Centrifuge the reaction following the mixing ensure that there is a clear supernatant for the capillary electrophoresis injector to draw from.
23. Replace the seal of the plate with a 96-well Septa for SeqStudio™ genetic analyzer, and load the plate into the Seq Studio for amplicon sequencing analysis (*see* **Note 10**).

3.3 Sanger Sequencing and Genotyping Analysis

1. Tap touchscreen of SeqStudio instrument to wake.
2. Login to Cloud Account. Enter Username: Enter Password: (*see* **Note 11**).
3. Ensure SeqStudio genetic analyzer cartridge has enough injections and is not expired.
4. Ensure SeqStudio genetic analyzer cathode buffer container is fresh and is not expired.
5. Open tray.
6. Load 96-well plate.
7. Select start.
8. Select setup new experiment.
9. Select sequencing.
10. Select long sequencing analysis and choose wells that will be injected into the Seq Studio cartridge.
11. Select run plate.
12. When run is completed, transfer .abi files to Thermo Fisher Cloud or local computer.

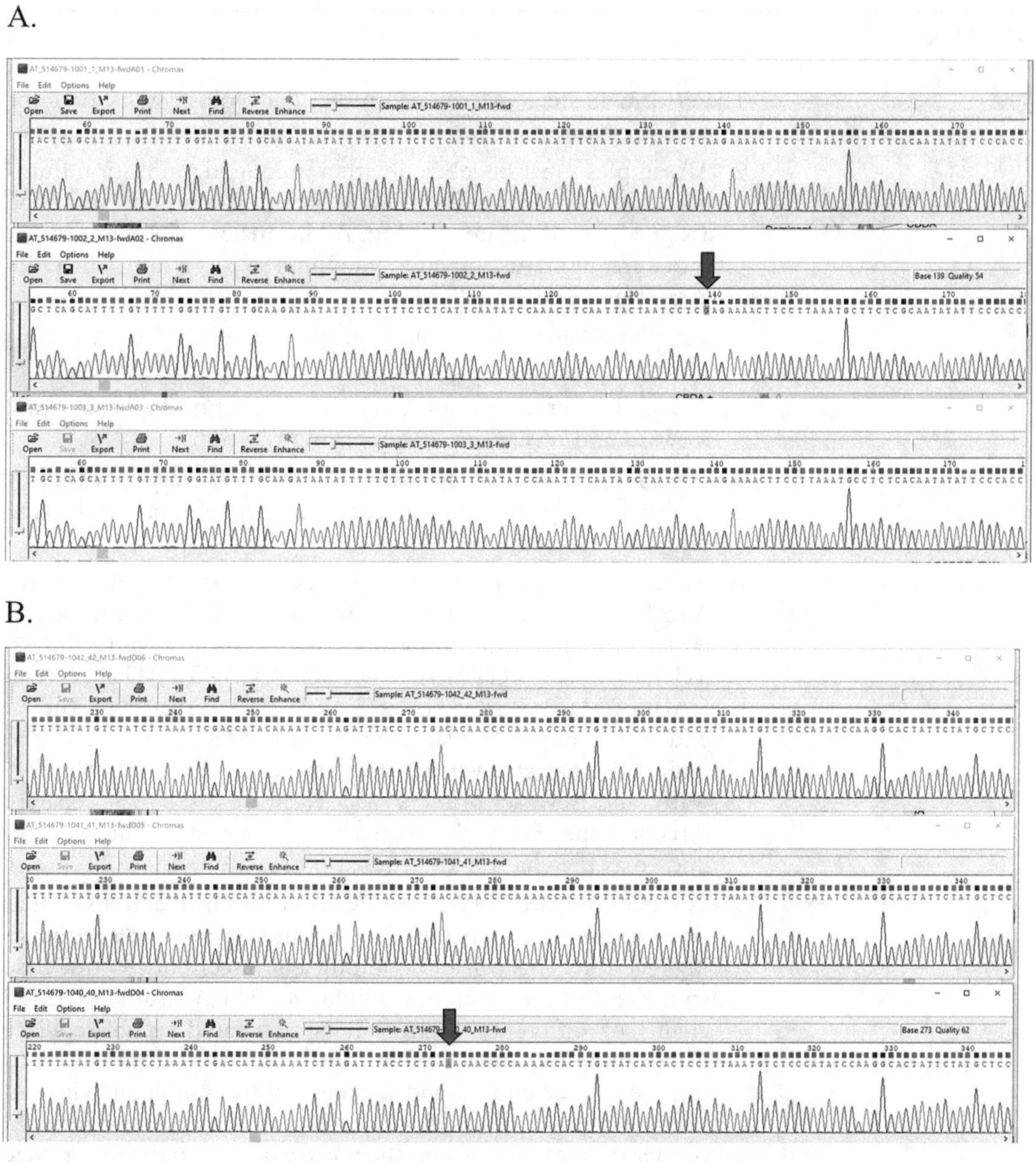

Fig. 1 Example sequencing chromatograms aligned to illustrate target genotype variation between sample sets (**a**) and (**b**) Arrows point to SNP genotype variation within each sample set

13. Open chromatogram or sequencing visualization software to view.
14. Refer to sequence references, analyze amplicon sequence, compare alleles for each genotype, and compare positions according to reference using sequence homology or alignment (Fig. 1) (*see* **Note 12**).

4 Notes

1. Perform all steps in a defined molecular biology workspace for each preparation step.
2. All samples must be labeled with a unique identification number, and an alphanumeric well position number must be assigned for all future labeling. Additionally, if FTA imprinted spots are light, use greater sampling size (~8 mm) in same extraction volume (100 μL).
3. Template volume for 1 reaction (μL) of about 4 ng/μL concentration of DNA preparation.
4. Test each template sample as at least one set of duplicates for a forward and reverse cycle sequencing reaction.
5. If testing a purified DNA sample from a DNEasy DNA kit or similar kit, skip the sample preparation steps and proceed to the direct PCR setup.
6. Annealing temperature for PCR reaction may vary on design. Aim for an annealing temperature of 60 °C for the annealed target sequence when ordering custom M13 tagged primers. M13 forward: TGTAAAACGACGGCCAGT; M13 reverse: CAGGAAACAGCTATGACC. Add M13 forward to the 5′ end of the target forward primer and M13 reverse to the 5′ end of the target reverse primer in a 5–3′ direction. The forward and reverse primers will each be used for separate direction-specific cycle sequencing reactions. This is also known as the chain-end termination reaction.
7. You have prepared two direct PCR master mix reactions so that you can prepare a separate forward and reverse cycle sequencing reaction. This will allow adding a florescent nucleotide base to the end of the extending amplicons. The mixture of labeled chains of different length will be separated using capillary electrophoresis, and the florescent ends will be measured to call the specific end-to-end sequence of the amplicon. This can only be done in a specific direction, so add only the forward primer to one reaction and the reverse primer to the other cycle sequencing reaction.
8. The SAM™ solution separates quickly so be sure to mix well and that the mixing occurs throughout the cleanup process using the Digital Vortex-Genie® 2.
9. Cover the plate with foil while sample plates are on the Digital Vortex-Genie® 2 as the reactions have fluorescent labels that should be protected from excessive light exposure.
10. Avoid aerosolizing amplicon products. Peel the film and add the capillary injection septa carefully to avoid splashes.

11. Create a login account with Thermo Fisher to utilize remote and cloud functions with the SeqStudio instrument.
12. Files will be written in .abi format, showing a base called sequence and measurement quality. Use chromatogram viewing software to perform a quality assessment of length and base called quality and accuracy.

Conflict of Interest

The authors have no conflicts of interest to declare that are relevant to the content of this chapter.

References

1. Gomes A, Korf B (2018) Genetic testing techniques. In: Pediatric cancer genetics, pp 47–64. https://doi.org/10.1016/b978-0-323-48555-5.00005-3
2. Endrullat C, Glokler J, Franke P, Frohme M (2016) Standardization and quality management in next-generation sequencing. Appl Transl Genom 10:2–9. https://doi.org/10.1016/j.atg.2016.06.001
3. Deng YM, Spirason N, Iannello P, Jelley L, Lau H, Barr IG (2015) A simplified Sanger sequencing method for full genome sequencing of multiple subtypes of human influenza A viruses. J Clin Virol 68:43–48. https://doi.org/10.1016/j.jcv.2015.04.019
4. Furutani S, Furutani N, Kawai Y, Nakayama A, Nagai H (2022) Rapid DNA sequencing technology based on the Sanger method for bacterial identification. Sensors (Basel) 22(6). https://doi.org/10.3390/s22062130
5. Loubet I, Caddoux L, Fontaine S, Michel S, Pernin F, Barres B, Le Corre V, Delye C (2021) A high diversity of mechanisms endows ALS-inhibiting herbicide resistance in the invasive common ragweed (Ambrosia artemisiifolia L.). Sci Rep 11(1):19904. https://doi.org/10.1038/s41598-021-99306-9
6. Deharvengt SJ, Petersen LM, Jung H-S, Tsongalis GJ (2020) Nucleic acid analysis in the clinical laboratory. In: Contemporary practice in clinical chemistry, pp 215–234. https://doi.org/10.1016/b978-0-12-815499-1.00013-2
7. Wang J, Wang W, Liu Y, Duo L, Huang L, Jiang X (2009) The method of single-nucleotide variations detection using capillary electrophoresis and molecular beacons. Mol Biol Rep 36(7):1903–1908. https://doi.org/10.1007/s11033-008-9397-y
8. Crossley BM, Bai J, Glaser A, Maes R, Porter E, Killian ML, Clement T, Toohey-Kurth K (2020) Guidelines for Sanger sequencing and molecular assay monitoring. J Vet Diagn Invest 32(6):767–775. https://doi.org/10.1177/1040638720905833
9. Seroussi E (2021) Estimating copy-number proportions: the comeback of Sanger sequencing. Genes (Basel) 12(2). https://doi.org/10.3390/genes12020283
10. Torres A, Pauli C, Givens R, Argyris J, Allen K, Monfort A, Gaudino RJ (2022) High-throughput methods to identify male Cannabis sativa using various genotyping methods. J Cannabis Res 4(1):57. https://doi.org/10.1186/s42238-022-00164-7
11. Al-Shuhaib MBS, Hashim HO (2023) Mastering DNA chromatogram analysis in Sanger sequencing for reliable clinical analysis. J Genet Eng Biotechnol 21(1):115. https://doi.org/10.1186/s43141-023-00587-6

Chapter 15

Long-Range Targeted Nanopore Ligation Sequencing Workflow for Targeted Amplicon Sequencing of Secondary Metabolite Gene

Anthony Torres and Reginald Gaudino

Abstract

This chapter presents protocols for advanced genotyping and genetic analysis in plant science, utilizing DNA extracted from FTA samples and leaf tissue. Targeted long range PCR is employed to uncover allelic and structural variations within genes of interest, while targeted long-read nanopore sequencing offers a comprehensive view of gene targets in their native and enriched forms. The protocol for nanopore sequencing encompasses DNA preparation, barcoding of up to 96 samples, adapter ligation, purification, and sequencing, culminating in the generation of raw data files (pod5 or fastq) suitable for further analysis. Together, these methods provide a powerful toolkit for plant diagnostics and breeding.

Key words Genotyping, Plant diagnostics, Agricultural biotechnology, FTA samples, Leaf tissue, Targeted long-range PCR, Long-read nanopore sequencing, DNA barcoding, MinION sequencer long-read DNA

1 Introduction

Single-molecule sequencing advancements using Oxford nanopore have enabled accurate determination of genotypes for various DNA targets and have created tangible accessibility using next-generation sequencing for applied molecular biology and genetic analysis [1, 2]. Long-read sequencing workflows for detecting haplotypes, structural variation, and SNPs have been developed into routinely used sequencing pipelines due to the improved accuracy, performance, and efficiency of the Oxford nanopore long-read sequencing MinION platform [3]. Template input from crude or purified genomic DNA prepared into purified PCR-generated amplicon is used for barcoding and adapter ligation as preparative unloading for the sequencing pore array. High-quality single-molecule long reads are unzipped base by base and read from the input native genomic DNA or targeted amplicon [4, 5]. A sequencing

Anthony Torres (ed.), *Genotyping: Methods and Protocols*, Methods in Molecular Biology, vol. 2943,
https://doi.org/10.1007/978-1-0716-4642-7_15,

experiment can run for up to 72 h, and prepared ligated template read recovery has routinely been reported in the millions of single molecules [2, 6, 7].

A focus on targeted sequencing is emphasized for this workflow to characterize a given gene/loci amplified from genomic DNA that is greater than 1 kb in amplicon fragment length, allowing an advantage of targeted nanopore sequencing over short-read sequences generated by Sanger or illumina-based sequencing methods [8]. Long-read amplicons from various samples to be genotyped and characterized are barcoded and pooled into a sequencing library. In a 96-barcode library, up to 96 samples are sequenced simultaneously, generating a high-throughput analysis of single-molecule amplicons that generate a read set for each amplicon/barcode. Sequence depth and read coverage depend on amplicon purity and quality [9, 10]. The recovered sequences from the targeted loci can be analyzed in real time by aligning and comparing them to a reference [5]. Sequence analysis of recovered targets enables the identification of genotypic features of a given long-read amplicon [4, 11]. Here we demonstrate an end-user workflow using the Oxford Nanopore Ligation Sequencing Kit prepared from >100 ng of targeted long-range PCR products.

2 Materials

2.1 Sample Preparation

1. Bunsen burner.
2. Calibrated p1, p100, p200 multichannel pipettes and p1, p200, p1000 single-channel pipettes.
3. 100–1000 μL nuclease-free aerosol filtered pipette tips.
4. 20–300 μL nuclease-free aerosol filtered pipette tips.
5. 1–100 μL nuclease-free aerosol filtered pipette tips.
6. 0.1–20 μL nuclease-free aerosol filtered pipette tips.
7. Stainless steel dissecting forceps.
8. 250 mL beaker.
9. Optional: plate piercer.
10. 1.5 mL centrifuge tubes.
11. Gloves.
12. 100% ethanol.
13. Nuclease-free water.
14. Extraction buffer solution: 0.25 M Tris–HCl, pH 8, 0.25% Triton-X.
15. Salt buffer solution: 5 M NaCl.

16. FTA plant saver cards.
17. Unicore punch kit, 2 mm.

2.2 Long-Range PCR

1. Standard thermocycler.
2. Mini plate spinner.
3. Gloves.
4. Clean designated space.
5. Clear 96-well plate or similar.
6. Sterile loading trough.
7. Long Amp oligonucleotides (1–10 kb targets) (*see* **Note 1**).
8. Long-range PCR mix.
9. Genotype-specific positive DNA for the positive control template.
10. AlumaSeal II sealing foils.

2.3 PCR Purification

1. DNA Clean and Concentrator MagBead Kit (*see* **Note 2**).
2. 96-well plate holder with magnet.
3. Molecular biology grade ethanol.

2.4 End-Prep Reaction

1. NEBNext® Ultra™ II End Repair/dA-Tailing Module (96 rxns) (New England Biolabs).

2.5 Native Barcoding and Pooling

1. Native Barcoding Expansion 96 (Oxford Nanopore).
2. LoBind DNA/RNA tubes.
3. Blunt/TA Ligase Master Mix.
4. AMPure XP, 5 mL.

2.6 Adapter Ligation and Library Purification

1. Ligation Sequencing Kit (Oxford Nanopore).
2. Adapter Mix II Expansion (Oxford Nanopore).
3. NEBNext® Ultra™ II Ligation Module (New England Biolabs).

2.7 Preparation of Loading Sequencing Library, Priming, and Loading Flow Cell

1. R9 Flow Cell (Oxford Nanopore).
2. DNA quantification kit (Qubit).
3. Ligation Sequencing Kit (Oxford Nanopore).
4. 50 mL conical tubes.
5. 1.5 mL centrifuge tube holder with magnet.
6. Inverting Hula Mixer.

3 Methods

3.1 Sample Preparation

1. Prepare the workstation on the lab bench by wiping the surface with alcohol where the preparation will be performed.
2. Prepare a clean PCR setup with clean 1.5 mL microcentrifuge tubes in an easily accessible area for sample preparation.
3. Sterilize forceps by dipping them in alcohol and flame sterilizing them or aseptically wiping excess alcohol until all alcohol has evaporated.
4. Remove a small sampling of raw tissue pieces or imprinted FTA card (~2–8 mm) with flame-sterilized forceps or alcohol-sterilized QIAGEN 2 mm UniCore punch tool, taking care not to touch the sample directly with gloves (*see* **Note 3**).
5. Carefully place the sample into a 1.5 mL microcentrifuge tube at the bottom of the well so that it does not contact the top of the tube.
6. Repeat **steps 2–5** with remaining lots/replicates until all samples are processed.
7. Fill the sterile liquid basin/trough with Extraction Buffer Solution using a dedicated single-channel or multichannel 30–300 μL multichannel pipettor loaded with clean tips to draw and dispense 200 μL of extraction buffer into each tube.
8. Repeat for all remaining samples, replacing tips and ensuring tubes are capped between steps.
9. Spin tubes down in microcentrifuge for 60 s to fully immerse all material in the extraction buffer.
10. Load the tube on a thermal incubation block and set to run for 10 min. Boil with lids sealed with tube lock to close.
    (a) ~95–100 °C 10 min 1×.
11. Following boil, add 8 μL 5 M NaCl and vortex.
12. Add 400 μL 100% ethanol and invert several times.
13. Place tubes at −80 °C overnight for 15 min or −20 °C to precipitate DNA.
14. Following ultra-cold precipitation, mark the position where the pellet will form on the tube and centrifuge 1.5 mL tubes for 30 min at 4 °C.
15. Remove tubes carefully, decant the supernatant, and retain the pellet.
    (a) Optional: Perform a 70% ethanol wash to remove additional residual waste.
16. Following wash, centrifuge 1.5 mL tubes at >10,000 × *g* or greater for 30 min at 4 °C.

17. Remove tubes carefully, decant the supernatant, and retain the pellet.
18. Allow the pellets to dry for 5–10 min at room temperature, then reconstitute the pellet in 60 μL of nuclease-free water.

3.2 Long-Range PCR Reaction Setup

1. Prepare a clean 1.5 mL centrifuge tubes for Master Mix and clean 96-Well PCR plate for targeted long-range PCR reactions.
2. Prepare a targeted long-range target PCR reactions with 20 μL final reactions with Qiagen ultrarun long-range master mix or similar and custom long-range primer mix for the target gene of interest (*see* **Note 1**) according to below:

   Reagent:

   4× Qiagen ultrarun long-range master mix: 5 μL.

   Long-range primer mix (10 μM): 1.5 μL.

   Nuclease-free water: 9.1 μL.
3. In the dedicated PCR space, use a 0.1–10 μL multichannel pipette, load 0.1–20 μL filtered pipette tips, and pipette 15 μL of the prepared reaction mix to each well of the plate, respectively, for each sample.
4. Change tips and prepare to load samples, use fresh tips for each new sample.
   (a) For all reactions tested with reconstituted DNA input, using a 0.1–10 μL multichannel pipette or single-channel pipette, draw and dispense 5 μL of reconstituted DNA input be sure tips are below surface of mix before dispensing, mix by pipetting up and down several time, then, eject tips after final mix.
   (b) For all reactions with purified genomic DNA, using a 0.1–10 μL multichannel pipette or single-channel pipette, draw and dispense 1 μL of kit (10 pg–1 μg/μL) purified genomic DNA from DNEasy or similar, ensure that tips are below surface of mix before dispensing, mix by pipetting up and down several times, then, eject tips after final mix.
5. After each sample well has been filled, obtain the control standards and aliquot 5 μL of positive control template, negative control template, and no template control into their respective wells.
6. Seal plate thoroughly with AlumaSeal sealing foil.
7. Spin plates for 10 s.
8. The plate is now ready for PCR.
9. Place the PCR plate in the thermocycler instrument and select and run the long-range PCR protocol:

(a) 93 °C for 3 min 1×.

(b) 93 °C for 30 s.

(c) 65 °C for (30 s per kb of Target-Each Gene of Interest Primer set may require optimized Tm) (*see* **Note 4**).

(d) Repeat steps **b** and **c** 35×.

(e) 72 °C for 10 min 1×.

(f) 4 °C hold.

3.3 Long-Range PCR Purification

1. After the long-range PCR reaction, pull the plate out of the thermocycler and allow it to acclimate to room temperature. Then, proceed to the PCR clean-up protocol using a PCR DNA clean-up kit with magnetic cleanup, such as a DNA Clean and Concentrator MagBead Kit or similar (*see* **Note 2**).
2. Add DNA binding solution supplied in the MagBead Kit to each PCR reaction at four times the volume of the PCR reaction. Mix by pipetting.
3. Add 20 μL of magnetic DNA binding solution to each mixture and allow to incubate at room temperature for 10 min with intermitted mixing.
4. Spin down the tube with centrifugation to pellet magnetic beads. Place the pellet on the magnet and allow it to remain undisturbed until the supernatant is clear. Incubate on the magnet for 5 min, enabling further pelleting.
5. Pipette off clear supernatant and add 500 μL of wash buffer supplied in the MagBead Kit.
6. Repeat **steps 4** and **5**.
7. Remove the second wash and ensure no residual wash remains. Allow the pellet to dry for about 45 s, but do not let it over dry and crack (*see* **Note 5**).
8. Remove the tube from the magnet, immediately add 15 μL nuclease-free water, and mix gently by flicking the tube.
9. Perform a quality control check using a Qubit Fluorometer to measure the DNA concentration from 1 μL of the purified long-range amplicon ready for end-prep reaction.

3.4 End-Prep Reaction

1. In 1.5 mL tubes or 96-well plates, prepare in nuclease-free water a working volume of 12.5 μL of >100 ng amplicon (>1 kb) each for end-prep reaction.
2. Mix 12.5 μL of each amplicon sample with 1.75 μL Ultra II End-prep Reaction Buffer and 0.75 μL Ultra II End-prep Enzyme mix for a final 15 μL end-prepped reaction.

3. Mix well by gently pipetting up and down several times, if used, seal the plate with an AlumnSeal foil and spin down using centrifugation.
4. Incubate end-prepped reaction in a two-step temperature hold of 20 °C for 5 min and 65 °C for 5 min.

3.5 Native Barcoding and Pooling

1. Thaw NEB Blunt/TA Ligase Master Mix on ice.
2. Thaw and maintain AMPure XP Beads at room temperature, mixing with vortexing or inversion immediately before use.
3. Thaw at room temperature the Oxford nanopore native barcodes intended for use (NB01-96) for the corresponding amplicons/samples to be tested. Each sample with be labeled with a unique barcode so to be multiplexed in the final sequencing library.
4. Mix Oxford nanopore native barcodes thoroughly and centrifuge before use. Please use one barcode per amplicon/sample.
5. In a 1.5 mL tubes or 96-well plate, prepare the barcoding reaction with 3 μL nuclease-free water mixed with 0.75 μL of end-prep reaction, 1.25 μL native barcode (NB01-96), and 5 μL NEB Blunt/TA ligase master mix added each for barcoding reaction (*see* **Note 6**).
6. Mix reactions at least ten times with slow and gentle pipetting technique. If used, seal the plate with an alumna seal foil and spin down using centrifugation.
7. Incubate barcoding reaction in a one-step temperature hold of 20 °C for 20 min.
8. To terminate the barcoding reaction, add 1 μL EDTA supplied in the kit.
9. Using a gentle pipetting technique, pool all the contents of the barcode samples into a single 1.5 mL Eppendorf LoBind microcentrifuge tube or similar.
10. After vortexing AMPure XP (AXP) Beads add 0.4× Volume of resuspended beads to pooled reaction.
11. Incubate tube inside empty 50 mL conical tube on inverting rotator mixer for 10 min at ambient temperature.
12. Spin down the tube with centrifugation to pellet magnetic beads. Place the pellet on a magnet and allow it to remain undisturbed until the supernatant is clear. Incubate on the magnet for 5 min, enabling further pelleting.
13. In nuclease-free water, prepare 2 mL of 80% ethanol solution.
14. Pipette off clear supernatant and add 700 μL of fresh 80% ethanol solution directly onto the packed pellet. Do not disturb the pellet.

15. Immediately discard the ethanol wash and repeat the wash step with another 700 μL of 80% ethanol.
16. Remove the second wash and ensure no residual ethanol remains. Allow the pellet to dry for about 45 s, but do not let it over dry and crack.
17. Remove the tube from the magnet and immediately add 35 mL of nuclease-free water, gently flicking the tube to mix.
18. Perform a quality control check by measuring the DNA concentration in 1 μL of the purified pooled barcoded library using a Qubit fluorometer.

3.6 Adapter Ligation and Library Purification

1. Thaw at room temperature the Oxford Nanopore Short Fragment Buffer (SFB) or Long Fragment Buffer (LFB) and Elution Buffer (EB) and NEBNext Quick Ligation Buffer.
2. Thaw the NEBNext Quick Ligation T4 DNA Ligase and Oxford Nanopore Native Adapter (NA) on ice (*see* **Note 7**).
3. In a single 1.5 mL Eppendorf LoBind microcentrifuge tube or similar, mix in this order: 30 μL of the Pooled Barcoded Library, 5 μL of Oxford Nanopore Native Adapter (NA), 10 μL NEBNext Quick Ligation Reaction Buffer, and finally 5 μL of NEBNext Quick T4 DNA Ligase. Mix reactions at least ten times with slow and gentle pipetting technique.
4. Incubate ligation reaction in a one-step temperature hold of 20 °C for 20 min.
5. To terminate ligation reaction, add 20 μL of resuspended AMPure (AXP) magnetic beads.
6. Mix gently by flicking the tube and incubate the tube inside empty 50 mL conical tube on inverting rotator mixer for 10 min at ambient temperature.
7. Spin down tube with centrifugation to pellet magnetic beads and place on magnet and allow pellet to remain undisturbed until supernatant is clear. Incubate on magnet for 5 min enabling further pelleting.
8. Pipette off clear supernatant and add either 125 μL of the long fragment buffer (LFB).
9. Mix gently by flicking the tube and spin down tube with centrifugation to pellet magnetic beads and place on magnet and allow pellet to remain undisturbed until supernatant is clear.
10. Repeat **steps 8** and **9**.
11. Remove second wash and ensure no residual ethanol remains. Allow the pellet to dry for about 45 s, but do not let the pellet over dry and crack.

12. Remove the tube from the magnet and immediately add 15 µL Oxford Nanopore Elution Buffer (EB) and mix by gently flicking the tube.
13. Mix by gently flicking the tube and incubate tube inside empty 50 mL conical tube on inverting rotator mixer for 10 min at ambient temperature.
14. Mix by gently flicking the tube and spin down tube with centrifugation to pellet magnetic beads and place on magnet and allow pellet to remain undisturbed until supernatant is clear.
15. Transfer the clear supernatant containing the ligated pooled library of amplicons to a single 1.5 mL Eppendorf LoBind microcentrifuge tube or similar tube.
16. Perform a quality control check using a Qubit Fluorometer to measure the DNA concentration from 1 µL of the purified adapter-ligated pooled library.

3.7 Preparation of Loading Sequencing Library, Priming, and Loading Flow Cell

1. At room temperature, thaw the Oxford Nanopore Sequencing Buffer (SB), Library Beads (LIB), Flow Cell Tether (FCT), and Flow Cell Flush (FCF) (*see* **Note 8**).
2. Prime the flow cell by adding a Flush Solution prepared with 1170 µL Oxford Nanopore Flow Cell Flush (FCF) and 30 µL Flow Cell Tether (FCT).
3. Open the MinION and insert the flow cell under the clip, inserting the pins into the MinION Flow Cell connection port.
4. Slide the flow cell priming port to the right to open the priming port.
5. Using a p1000 pipette set to 200 µL, insert the pipette tip into the port and remove air bubbles from the port by drawing a small volume (20–30 µL) until yellow storage buffer enters the tip from the priming port creating a fluid stream. Be sure not to overdraw and expose the sensor array.
6. Load 800 µL of the prepared Oxford Nanopore Flush Solution (from **step 2**) into the priming port without adding any air bubbles and wait for 5 min to allow priming. Prepare the loading reaction with ligated library.
7. Dilute the library to 10–20 fmol to 12 µL working volume.
8. In a single 1.5 mL Eppendorf LoBind microcentrifuge tube or similar, mix in this order: 37.5 µL Oxford Nanopore Sequencing Buffer (SB), 25.5 µL library beads (LIB), and 12 µL of pooled library of amplicons for a total of 75 µL library for loading.
9. Open the Oxford Nanopore SpotON sample port gently by lifting it from its holding position, and pipette and load 200 µL

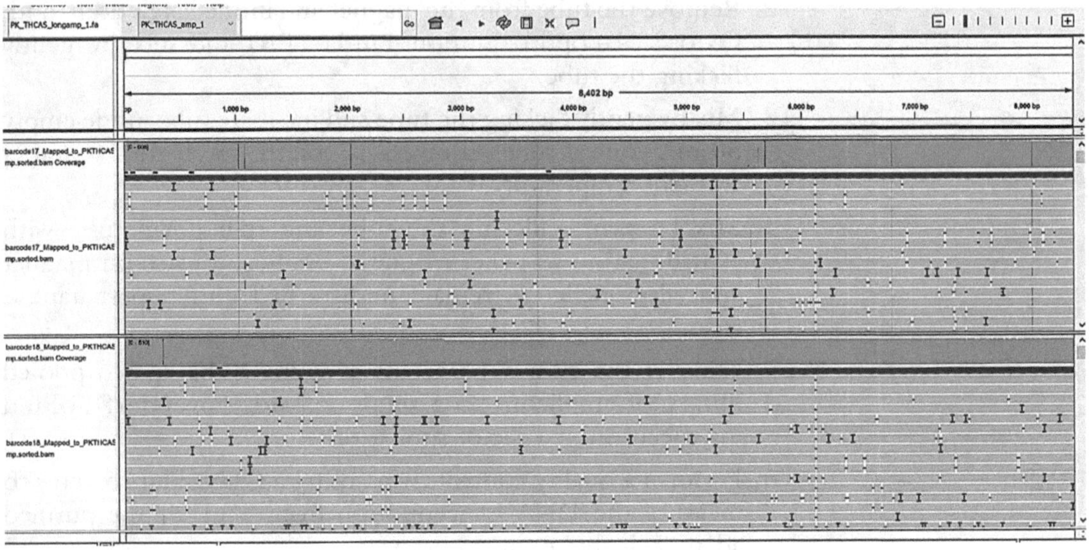

Fig. 1 An example alignment of two 8402 base pair amplicons generated via nanopore sequencing. These amplicons target the same gene of interest locus in two different genotypes. The mapped reads provide over 100× coverage, allowing for high-confidence identification of single nucleotide polymorphisms (SNPs) that distinguish the genotypes (reads highlighted in green, orange, red, and blue)

of Oxford Nanopore Flush Solution into the flow cell priming port (NOT the SpotON sample port). Droplets may form on the SpotON sample port as Flush solution is added, but allow fluid dynamic movement and please avoid adding any air bubbles (*see* **Note 9**).

10. To the SpotON sample port, add and load onto flow cell array in a drop-wise fashion 75 μL of the library. Ensure the droplet flows into the sample port before adding additional drops. Again, avoid adding any air bubbles as they will damage the array (*see* **Note 10**).
11. After the library has been loaded, replace the SpotON sample port cover, making sure that the tip end fits snugly into the SpotON sample port, and also close the priming port cover.
12. Close the device lid and set up a sequencing run using Oxford Nanopore MinKNOW software.
13. Following the sequencing run and QC, map high-quality reads to reference for genotyping analysis (Fig. 1).

4 Notes

1. Oligonucleotides must be well designed and specific to produce one (1–10 kb) long-range amplicon product.

2. The quickest way to clean up PCR products is via a magnetic bead clean-up kit, but other PCR clean-up protocols, such as column-based purification, can also be used.
3. A 2 mm hole punch taken from leaf material or FTA printed samples can be used to input material into crude extraction, producing a lysate containing nucleic acids for template input into long-range PCR. The size of the hole punch may be adjusted for higher DNA yields.
4. The annealing temperature for the designed oligonucleotide prepared for the long-range PCR primer mix should be at least 60 °C.
5. Ensure that all residual ethanol is removed after the wash step during PCR product cleanup. The amplicon product must be free of solvent contamination. Also, pipetting should be minimized during handling, especially for longer PCR products (for example >5 kb).
6. Each purified amplicon product will be assigned a unique barcode included in the native barcode expansion sequencing kit (Oxford Nanopore).
7. Ensure the native adapter is used with amplicons prepared with native barcodes. This is critical for preparing a successful targeted amplicon sequencing library.
8. The library beads settle quickly ensure they are mixed well prior to pipetting by gently flicking the tube.
9. It is critical to maintain the fluid dynamics and not to introduce any air bubbles into the flow cell. When adding a flushing buffer to the flushing port of the flow cell, directly insert the pipette tip connecting into the flushing port and slowly draw ~20 μL volume of the buffer from the flushing port into the pipette tip. A tiny liquid droplet may form at the flushing port, and directly insert the pipette tip into the droplet, making direct contact with the port and pipette slowly 200 μL of flushing solution into the flow cell. When flushing the flow cell, repeat this process. Please ensure no air bubbles are introduced into the flow cell.
10. A sequencing run can be paused to load more libraries following a flow cell flush as long as active pores are available for sequencing.

Conflict of Interest

The authors have no conflicts of interest to declare that are relevant to the content of this chapter.

References

1. Deamer D, Akeson M, Branton D (2016) Three decades of nanopore sequencing. Nat Biotechnol 34(5):518–524. https://doi.org/10.1038/nbt.3423
2. Kono N, Arakawa K (2019) Nanopore sequencing: review of potential applications in functional genomics. Dev Growth Differ 61(5):316–326. https://doi.org/10.1111/dgd.12608
3. Malmberg MM, Spangenberg GC, Daetwyler HD, Cogan NOI (2019) Assessment of low-coverage nanopore long read sequencing for SNP genotyping in doubled haploid canola (Brassica napus L.). Sci Rep 9(1):8688. https://doi.org/10.1038/s41598-019-45131-0
4. Whitford W, Hawkins V, Moodley KS, Grant MJ, Lehnert K, Snell RG, Jacobsen JC (2022) Proof of concept for multiplex amplicon sequencing for mutation identification using the MinION nanopore sequencer. Sci Rep 12(1):8572. https://doi.org/10.1038/s41598-022-12613-7
5. Schmidt J, Berghaus S, Blessing F, Herbeck H, Blessing J, Schierack P, Rodiger S, Roggenbuck D, Wenzel F (2022) Genotyping of familial Mediterranean fever gene (MEFV)-single nucleotide polymorphism—comparison of Nanopore with conventional Sanger sequencing. PLoS One 17(3):e0265622. https://doi.org/10.1371/journal.pone.0265622
6. Ng AWT, McClurg DP, Wesley B, Zamani SA, Black E, Miremadi A, Giger O, Hoopen RT, Devonshire G, Redmond AM, Grehan N, Jammula S, Blasko A, Li X, Aparicio S, Tavare S, Oesophageal Cancer Clinical and Molecular Stratification (OCCAMS) Consortium, Nowicki-Osuch K, Fitzgerald RC (2024) Author Correction: Disentangling oncogenic amplicons in esophageal adenocarcinoma. Nat Commun 15(1):4533. https://doi.org/10.1038/s41467-024-48912-y
7. Gurgul A, Jasielczuk I, Szmatola T, Sawicki S, Semik-Gurgul E, Dlugosz B, Bugno-Poniewierska M (2023) Application of Nanopore sequencing for high throughput genotyping in horses. Animals (Basel) 13(13). https://doi.org/10.3390/ani13132227
8. McNaughton AL, Roberts HE, Bonsall D, de Cesare M, Mokaya J, Lumley SF, Golubchik T, Piazza P, Martin JB, de Lara C, Brown A, Ansari MA, Bowden R, Barnes E, Matthews PC (2019) Illumina and Nanopore methods for whole genome sequencing of hepatitis B virus (HBV). Sci Rep 9(1):7081. https://doi.org/10.1038/s41598-019-43524-9
9. Delahaye C, Nicolas J (2021) Sequencing DNA with nanopores: troubles and biases. PLoS One 16(10):e0257521. https://doi.org/10.1371/journal.pone.0257521
10. Groen K, van Nieuwkoop S, Bestebroer TM, Fraaij PL, Fouchier RAM, van den Hoogen BG (2021) Whole genome sequencing of human metapneumoviruses from clinical specimens using MinION nanopore technology. Virus Res 302:198490. https://doi.org/10.1016/j.virusres.2021.198490
11. Yeo D, Chan AHJ, Hiong KC, Ong J, Ng JY, Lim JM, Zhang W, Lim SR, Fernandez CJ, Wong AM, Lee BPY, Khoo MDY, Cheng TXW, Lim BTM, Yeo HHT, Tan MMQ, Sng WBG, Adam SS, Ang WF, How CB, Xie R, Wasser SK, Finch KN, Loo AHB, Yap HH, Leong CC, Er KBH (2023) Uncovering the magnitude of African pangolin poaching with extensive nanopore DNA genotyping of seized scales. Conserv Biol 38:e14162. https://doi.org/10.1111/cobi.14162

Chapter 16

Double Digest Restriction-Site Associated DNA Sequencing (ddRAD-Seq)

Zenaida V. Magbanua, Chuan-Yu Hsu, Olga Pechanova, Mark Arick II, and Daniel G. Peterson

Abstract

ddRAD-Seq is a reduced representation sequencing technique that results in sequence datasets that can be compared and used to identify SNPs. We present an improved ddRAD-Seq protocol that increases efficiency and decreases the time to complete a ddRAD-Seq experiment. It utilizes selected restriction enzyme digestion fragments, quick acting ligases that are neutral with the restriction enzyme buffer eliminating buffer exchange steps, and adapters designed to be compatible with Illumina index primers. Enzyme deactivation steps are eliminated, library amplification and barcoding are completed in one PCR step, highly-efficient and precise size selection with BluePippin system and cleanup steps using magnetic beads are consolidated at the end of the library generation step. The SNPs that we identified using this streamlined protocol were validated in population and evolutionary studies of cotton (plant) (Magbanua et al., Anal Biochem 662:115001, https://doi.org/10.1016/j.ab.2022.115001, 2023) and rohu carp (animal) (Arick et al., G3 (Bethesda) 13, https://doi.org/10.1093/g3journal/jkad009, 2023).

Key words Reduced genome representation, Next-generation sequencing, Restriction enzyme digestion, SNP identification

1 Introduction

Double digest restriction-site associated DNA-Seq (ddRAD-Seq) is a reduced genome representation genotyping method that utilizes next-generation sequencing (NGS) technologies [3]. In general, it starts with high molecular weight genomic DNA and involves restriction enzyme digestion, adapter ligation, barcoding, library amplification, and NGS. A well-designed and planned experiment will provide useful data with a minimum input of resources and time. To this end, several modifications have been put forth to advance the technique such as using a divergent-Y in the second adapter [4], using a single divergent-Y adapter and attaching the barcodes during PCR [5], and optimization of the reagents

Anthony Torres (ed.), *Genotyping: Methods and Protocols*, Methods in Molecular Biology, vol. 2943,
https://doi.org/10.1007/978-1-0716-4642-7_16,

including thoughtful design of the adapters and procedures to increase the efficiency and decrease the time to complete the experiment [1]. A comprehensive review of early protocols was discussed by Andrews et al. [6]. Here, we demonstrate a streamlined ddRAD-Seq method that builds upon existing work, uses mostly commercially available reagents, eliminates steps like buffer exchanges and enzyme deactivation, and consolidates all the size selection and cleanup steps to just one after library construction. The generated libraries are designed to be sequenced using the Illumina platform. Sequence analysis will depend heavily on the objective of the experiment, and the system being studied. We illustrated this with two examples: (1) we compared the phylogenetic results of whole genome resequencing and ddRAD-seq variant calling in cotton (*Gossypium* A-genome) [1] and (2) we found a possible sex-determination mechanism of the rohu carp using the presence/absence of mapped reads [2].

2 Materials

2.1 In-Silico Digestion and Enzyme Pair Selection

Computer with a recent version of Windows, Linux, or Mac.

2.2 Double Digestion of Genomic DNA

1. High-molecular weight genomic DNA that has been verified on a 0.8% w/v agarose gel (*see* **Note 1**).
2. Frequent cutter restriction enzyme (FCRE).
3. Rare cutter restriction enzyme (RCRE).
4. PCR 8-tube strips.
5. Nuclease-free water.
6. Milli-Q water.
7. Thermocycler.
8. Agarose powder.
9. Gel running buffer: Tris-acetate-EDTA (50× TAE Buffer); Store at room temperature.
10. Ultra Pure™ Ethidium Bromide, 10 mg/mL. Store in a 4 °C refrigerator.
11. Gel loading buffer: 10× Invitrogen™ BlueJuice™ Gel Loading Buffer; store in a 4 °C refrigerator.
12. Molecular weight marker. 1 kb Plus DNA Ladder; store in a 4 °C refrigerator.
13. Gel electrophoresis apparatus.
14. Gel imaging system.

2.3 Ligation of Adapters

1. Double digested genomic DNA.
2. FCRE adapter duplex (*see* **Note 2**).
   5′-5Phos/FCRE cutting site CTGTCTCTTATACA CATCTCCGAGCCCACGAGAC-3′ and 5′-GTCTCGTGG GCTCGGAGATGTGTATAAGAGACAG FCRE cutting site-3′. Store solutions in a −20 °C freezer.
3. RCRE adapter duplex (*see* **Note 2**).
   5′-5Phos/RCRE cutting site CTGTCTCTTATACA CATCTGACGCTGCCGACGA-3′ and 5′-TCGTCGGCAG CGTCAGATGTGTATAAGAGACAG RCRE cutting site-3′. Store solutions in a −20 °C freezer.
4. Quick ligase. Quick Ligation Kit. New England Biolabs (*see* **Note 3**). Store solutions in a −20 °C freezer.
5. 5× Ligase buffer. NEBNext Quick Ligation Reaction Buffer. New England Biolabs (*see* **Note 3**). Store solutions in a −20 °C freezer.

2.4 Construction of Sequencing Library: Barcoding via PCR and Pooling of Samples

1. PCR 8-tube strips.
2. Nextera XT Index Kits v2. Illumina Set A, Set B, Set C, or Set D. Store in a −20 °C freezer.
3. Minicentrifuge for 8-tube strips.
4. Vortex mixer.
5. Thermocycler.
6. Q5 High-Fidelity 2× Master Mix. New England BioLab.
7. Nuclease-free water.

2.5 Cleanup and Size Selection of Library

1. Agencourt AMPure XP beads: Beckman Coulter. Store at 4 °C refrigerator.
2. Magnetic stand for 1.5-mL microcentrifuge tubes.
3. Microcentrifuge.
4. Vortex mixer.
5. Microcentrifuge tube rotator.
6. Absolute ethanol (200 Proof). Store at room temperature.
7. 10 mM Tris–HCl (pH 8.5). Store at room temperature.
8. Sage Science BluePippin system.
9. BluePippin 2% Agarose Gel Cassette (100–600 bp). Sage Science.

2.6 Validation of Library

1. Qubit fluorometer.
2. Qubit dsDNA HS Assay Kit.
3. Qubit assay tubes.
4. Microcentrifuge.

5. Vortex mixer.
6. Agilent Bioanalyzer 2100 system.
7. Agilent chip priming station.
8. IKA vortex mixer.
9. Agilent DNA 1000 Kit. Agilent Technologies.

3 Methods

3.1 In-Silico Digestion and Enzyme Pair Selection

1. Download the latest egads release from https://github.com/IGBB/egads/releases.
2. Download genome. Use the closest species with a chromosome level genome assembly. NCBI Refseq or EMBL-EBI Ensembl are good places to start.
3. Find list of restriction enzymes. Any enzyme found in REBASE (the restriction enzyme database) can be used. The available enzymes can be found in various lists at http://rebase.neb.com/rebase/rebase.charts.html.
4. Run egads. Rare (--rare) and frequent (--freq) cutting enzymes are specified by a comma delimited list of enzyme names. The genome (--genome) must be a valid fasta file (can be gzip compressed). The --html argument specifies the HTML report output file. Below is an example of running egads on the *G. longicalyx* genome using three rare and four frequent cutting restriction enzymes.

```
./egads --rare NsiI,HindII,PstI \
--freq Sau3AI,BfaI,MspI,HinP1I \
--genome longicalyx.fasta.gz \
--html longicalyx.egads.html
```

5. Use the HTML report to select candidate pairs. The left column of the table shows the compatibility of buffers reported in REBASE. The ridge plot shows the size distribution of valid fragments (i.e., fragments between the two enzymes). The gel image shows the length distribution of all fragments. Adjusting the selected size by resizing the gray box in the ridge plot updates both the table and the ridge plot. Select enzyme pairs that have compatible buffers, digest the genome well (both the ridge plot and expected gel image do not show many fragments above or below the target range), and retain around 5–10% of the genome (*see* **Note 4**).

3.2 Double Digestion of Genomic DNA

1. Set up the reactions in 8-tube strip PCR tubes (*see* **Note 5**). For each sample:

   Genomic DNA: 350 ng (calculate the volume based on concentration).
   FCRE: 1 μL.
   RCRE: 1 μL.
   10× Cut Smart Buffer: 2.5 μL.
   Nuclease-free water to make 25 μL total volume.

   Incubate the tubes in a thermocycler set at 37 °C for 2 h. At completion, proceed to **step 2** and after taking 5 μL from each tube, either store the remaining digested products at −20 °C freezer or keep them on ice if immediately proceeding to Subheading 3.3, **step 1** after gel electrophoresis.

2. Transfer the following into another 8-strip tube for gel electrophoresis:

   10× Gel loading buffer: 1 μL.
   Double-digestion reaction: 5 μL.
   Keep on ice while doing **step 3**.

3. Prepare 1× TAE by diluting the 50× stock with Milli-Q water (*see* **Note 6**). Use 1× TAE to make a 1.5% w/v agarose gel. Heat the agarose powder in 1× TAE in a microwave and add ETBr to a final concentration of 0.2–0.5 μg/mL before casting. Allow the agarose gel to polymerize before the next step. This may take more than 1 h.

4. Load all 6 μL of each digestion reaction in the well of the 1.5% w/v agarose gel and run the gel in 1× TAE buffer until the bromophenol blue dye makes its way halfway through the length of the gel (*see* **Note 7**).

3.3 Ligation of Adapters

1. Set up ligation reactions in the same 8-tube strip PCR tubes. Add to each tube (contains the remaining 20 μL from Subheading 3.2, **step 1**):

   Quick T4 DNA Ligase: 1 μL.
   0.1 μM RCRE Adapters: 1 μL.
   10 μM FCRE Adapters: 1 μL.
   5× ligation buffer: 6 μL.
   Nuclease-free water: 1 μL.
   Total volume: 30 μL.

   Incubate the completed ligation at room temperature for 10 min and either proceed to the next step or store the reactions at −20 °C until ready for the next step.

3.4 Construction of Sequencing Library: Barcoding via PCR and Pooling of Samples

1. Assign N7 and S5 barcode IDs to each sample (*see* **Note 8**).
2. Label 8-tube strip PCR tubes to use (*see* **Note 4**).
3. Set up each PCR barcoding reaction mix as below:
   Adapter-ligated DNA (from Subheading 3.3, **step 1**): 5.0 μL.
   Nextera XT N7 (i7) primer: 0.75 μL.
   Nextera XT S5 (i5) primer: 0.75 μL.
   Q5 High-Fidelity 2× Master Mix: 7.5 μL.
   Nuclease-free water: 1.0 μL.
   Final volume: 15 μL.
4. Store the 8-tube strip(s) containing the remaining adapter-ligated DNA samples at −20 °C in a freezer (tubes from Subheading 3.3, **step 1**).
5. Gently vortex the sample tubes (from **step 3**) and briefly spin down.
6. Place the 8-tube strip(s) (from **step 3**) in a thermocycler with the following program:
   98 °C for 30 s.
   98 °C for 10 s.
   65 °C for 30 s: 14 cycles.
   72 °C for 30 s.
   72 °C for 5 min.
   4 °C: Hold.
7. Pool the library samples (up to 96 samples) assigned to sequence on the same lane together by pipetting 7.5 μL from each barcoded library into a 1.5-mL microcentrifuge tube. Store the remaining barcoded library products in a −20 °C freezer.

3.5 Cleanup and Size Selection of Library

3.5.1 Library Pool Cleanup

1. Equilibrate the AMPure XP beads to room temperature for at least 30 min.
2. Prepare 80% ethanol in water by pipetting 4 mL of absolute ethanol and 1 mL of nuclease-free water into a 15-mL centrifuge tube and mixing well (*see* **Note 9**).
3. Vortex equilibrated AMPure XP beads for 1 min.
4. Add 384 μL (~0.8× of total volume of library pool) of resuspended beads (*see* **Notes 10–12**) to the 480 μL of library pool if pooling 96 independent barcoded libraries (*see* **Note 13**).
5. Mix well by gentle vortexing.
6. Place the tube on a rotator and rotate at room temperature for 5 min.
7. Place the tube in a microcentrifuge and spin with the hinge of the tubes oriented outward for 30 s.

8. Place the tubes on the magnetic stand such that the hinges are adjacent to the magnet. Allow to stand until the solution is clear.
9. Remove and discard the supernatant and wash the beads by adding 200 μL of 80% ethanol while keeping the tube on the magnetic stand.
10. Incubate at room temperature for 30 s.
11. Carefully remove and discard the supernatant without touching the magnetic beads.
12. Repeat the wash step (**steps 9–11**).
13. Place the tube in the microcentrifuge and spin for 30 s.
14. Place the tube back on the magnetic stand.
15. Carefully remove remaining ethanol with a P10 (or P20) pipette.
16. Air dry the magnetic beads (but not to the point of cracking) while still on the magnetic stand.
17. Add 32 μL of nuclease-free water or EB buffer (10 mM Tris–HCl pH 8.5) to the tube to elute the library pool.
18. Mix well by gentle vortexing.
19. Place the tube on a rotator and rotate at room temperature for 10 min.
20. Place the tube in the microcentrifuge and spin with the hinge of the tubes oriented outward for 30 s.
21. Place the tube on the magnetic stand such that the hinges are adjacent to the magnet. Allow to stand until the solution is clear.
22. Transfer the supernatant containing the purified library pool (at least 30 μL) to a new microcentrifuge tube.

3.5.2 Size Selection of Library Samples (Pools) with Sage Science BluePippin System

1. Go to "Protocol Editor" tab in the BluePippin software.
2. Click "Cassette" folder and select "2% DF Marker M1" in cassette definition.
3. Enter "1" in the "Reference Lane" field if deciding to load the M1 DNA marker into "Lane 1."
4. Click the "APPLY REFERENCE TO ALL LANES" button.
5. Select "Range" collection mode for each other lane and enter "350" for "BP Start" and "550" for "BP End" in the size selection parameters.
6. Press "Save As" to name and save the protocol.
7. Go to "Main" tab and the protocol name saved in the previous step.

8. Open the lid and place the calibration fixture on the optical nest to cover the LED detectors.
9. Close the lid and press "CALIBRATE" on the control panel.
10. Press "EXIT" once the message "Calibration OK" is shown in "Calibration Status" field and open the lid and remove the fixture.
11. Remove a gel cassette from the foil packaging and inspect the buffer levels in all buffer reservoirs. The buffer levels should be nearly full. Fill with spare buffer if less than full.
12. Inspect the agarose column and bubbles for each channel.
13. Open the lid and place the cassette into the optical nest, then remove the white tabbed adhesive strips from the cassette.
14. Remove the buffer from all the elution modules and add 40 μL of fresh electrophoresis buffer to each elution module and seal the elution wells with the adhesive tape strip.
15. Close the lid and press "TEST" to perform the continuity test.
16. Press "RETURN" once the "PASS" message is shown.
17. Prepare each sample by adding 10 μL of loading solution to 30 μL of purified library sample.
18. Mix well by gentle vortexing and follow by brief centrifugation.
19. Open the lid and remove 40 μL of buffer from each sample well and load 40 μL of DNA marker M1 in lane 1 well and 40 μL of each sample prepared in previous steps in the remaining wells.
20. Close the lid and press "START" on the control panel.
21. Open the lid once the run including separation and elution is completed.
22. Remove the adhesive tape strip from the elution wells and transfer the eluent (about 40 μL) from each sample elution well to a 1.5-mL centrifuge tube.

3.5.3 Size-Selected (Final) Library Pool Sample Cleanup

1. Equilibrate the AMPure XP beads to room temperature for at least 30 min.
2. Prepare 80% ethanol in water by pipetting 4 mL of absolute ethanol and 1 mL of nuclease-free water into a 15-mL centrifuge tube and mixing well (*see* **Note 9**).
3. Vortex equilibrated AMPure XP beads for 1 min.
4. Add 60 μL (~1.5× of 40 μL of size-selected library pool sample) of resuspended beads.
5. Mix well by gentle vortexing.
6. Place the tube on a rotator and rotate at room temperature for 5 min.

7. Place the tube in a microcentrifuge and spin with the hinge of the tubes oriented outward for 30 s.
8. Place the tubes on the magnetic stand such that the hinges are adjacent to the magnet. Allow to stand until the solution is clear.
9. Remove and discard the supernatant and wash the beads by adding 200 μL of 80% ethanol while keeping the tube on the magnetic stand.
10. Incubate at room temperature for 30 s.
11. Carefully remove and discard the supernatant without touching the magnetic beads.
12. Repeat the wash step (**steps 9–11**).
13. Place the tube in the microcentrifuge and spin for 30 s.
14. Place the tube back on the magnetic stand.
15. Carefully remove remaining ethanol with a P10 (or P20) pipette.
16. Air dry the magnetic beads (but not to the point of cracking) while still on the magnetic stand.
17. Add 17 μL of nuclease-free water or EB buffer (10 mM Tris–HCl pH 8.5) to the tube to elute the library pool.
18. Mix well by gentle vortexing.
19. Place the tube on a rotator and rotate at room temperature for 10 min.
20. Place the tube in the microcentrifuge and spin with the hinge of the tubes oriented outward for 30 s.
21. Place the tube on the magnetic stand such that the hinges are adjacent to the magnet. Allow to stand until the solution is clear.
22. Transfer the supernatant containing the purified final library pool (at least 15 μL) to a new microcentrifuge tube.
23. Aliquot 2 μL of final library pool for validation and store the remaining library pool sample in a freezer at −20 °C.

3.6 Validation of Libraries

3.6.1 Quantify Library Samples with a Qubit Fluorometer

1. Prepare Qubit dsDNA HS assay working solution by pipetting 1.99 mL of Qubit dsDNA HS buffer and 10 μL of Qubit dsDNA HS reagent in a 2-mL microcentrifuge tube (*see* **Note 14**).
2. Vortex the microcentrifuge tube to mix well.
3. Prepare assay mixes for each library in the Qubit assay tube as follows:

   Qubit dsDNA HS Standard 1: 190 μL of working solution + 10 μL of Standard 1.

Qubit dsDNA HS Standard 2: 190 μL of working solution + 10 μL of Standard 2.

Library samples: 199 μL of working solution + 1 μL of Library sample.

4. Vortex the assay tubes for 2–3 s and centrifuge for 30 s.
5. Incubate the assay tubes at room temperature for 2 min.
6. Read the assay tubes with Qubit fluorometer and record the concentration of each sample.

3.6.2 Analyze the Size of Library Samples with Agilent Bioanalyzer 2100

1. Equilibrate the DNA dye concentrate and DNA gel matrix to room temperature for 30 min.
2. Place a new syringe in the chip priming station.
3. Adjust the base plate of chip priming station to Position C.
4. Adjust the syringe chip to the lowest position.
5. Vortex the DNA dye concentrate and pipet 25 μL of DNA dye concentrate into a DNA gel matrix vial.
6. Vortex the gel/dye mix and spin down.
7. Transfer the solution into a spin filter and centrifuge at 2240 × *g* for 15 min at room temperature.
8. Use the filtered gel/dye mix for DNA chip assay and store the remaining solution in a 4 °C refrigerator and protect from light.
9. Pipet 9 μL of gel-dye mix into a well on a DNA assay chip marked as "G" with white font color and black background color.
10. Pull the plunger of syringe at the position of 1 mL and open the chip priming station.
11. Place the DNA assay chip on the base plate of chip priming station and close the chip priming station.
12. Press the plunger of syringe until held by the clip and wait for 60 s to release the clip.
13. Wait for 5 s and pull the plunger slowly back to the 1 mL position.
14. Open the chip priming station and remove the DNA assay chip from the chip priming station and place it on the bench top.
15. Pipet 9 μL of gel-dye mix into other two wells (9 μL each) on a DNA assay chip marked as "G" with black font color and no background color.
16. Pipet 5 μL of DNA marker into all 12 sample wells and the ladder well.
17. Pipet 1 μL of DNA ladder into the ladder well.

18. Pipet 1 μL of each library sample (with the concentration of 10–50 ng/μL) into the sample wells, and pipet 1 μL of DNA marker (or nuclease-free water) into each unused sample well.
19. Place the DNA assay chip on the IKA vortex mixer and vortex at 2400 rpm for 1 min.
20. Run the assay chip in Agilent Bioanalyzer 2100 system within 5 min.
21. Determine the mean fragment size for each library sample using the Bioanalyzer output.

3.6.3 Calculate Molarity Concentration of Library Sample

1. Use the formula (*see* **Note 15**):

   $$\text{nM} = [(\text{concentration in ng}/\mu\text{L})/(\text{library size in bp} \times 660\ \text{g per mole})] \times 10^6$$

   where concentration is measured by the Qubit fluorometer and the library mean size is determined by the Agilent Bioanalyzer.
2. Prepare the library sample with the required concentration (in nM) for sequencing.

4 Notes

1. After gel electrophoresis, the strongest band should be at or slightly above 25 kb with minimal shearing throughout the length of the well (*see* Fig. 1).

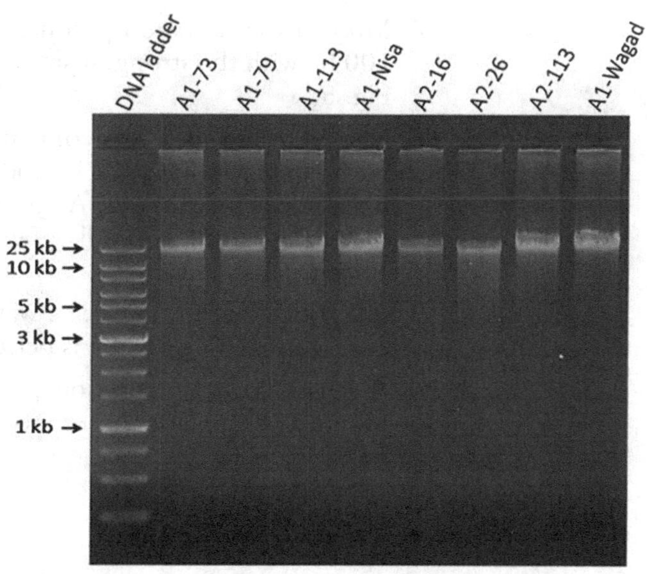

Fig. 1 Agarose gel image of extracted nuclear genomic DNA from ground leaves

2. The adapter pairs are complementary and are synthesized, annealed, and HPLC-purified after annealing by Integrated DNA Technologies (https://www.idtdna.com/pages/products/custom-dna-rna/dna-oligos/custom-dna-oligos). The green highlights represent the restriction enzyme cutting sites. Design the oligonucleotides such that the overhangs arising from the restriction enzyme digestion are in the correct end. For example, *Hin*dIII generates the cut 5′-A/AGCTT-3′ to yield a 5′ overhang. Consequently, one of the pairs of oligonucleotides should start (5′–3′) with AGCTT while the other pair should end with A (5′–3′). The yellow highlights represent the PCR priming site during the library construction (also the barcoding) step. Dissolve the synthesized, annealed adapters with 10 mM Tris–HCl (pH 8.5) to make 100 μM stocks and dilute the stock with nuclease-free water to make 0.1 μM (RCRE) or 10 μM (FCRE adapters) working solutions.
3. The Quick Ligation Kit comes with a 2× ligation buffer but we prefer to use a 5× concentration to minimize the reaction volume.
4. The egads report for *G. longicalyx* can be found at https://htmlpreview.github.io/?https://github.com/IGBB/egads/blob/main/example.html. Examples from that report are shown in Figs. 2, 3, and 4.
5. We advise working with groups of 8 samples, up to a maximum of 96 samples each time. However, if there are less than eight samples, regular PCR tubes may be used instead.
6. 1× TAE may be prepared in large quantities, stored at room temperature, and used as needed for weeks or months.
7. Most of the digestion products should be between 3 kb and 100 bp with the strongest smear between 1 kb and 300 bp (*see* Fig. 5).
8. The Nextera Index Kits come in four sets, each set providing 96 unique dual-indexed barcodes, for a total of 384 unique dual-indexed barcodes. A group of 96 samples each with unique barcodes will be sequenced together in 1 sequencing lane.
9. 80% v/v ethanol solution in water needs to be prepared fresh every time the procedure is performed.
10. A left-sided bead selection procedure will be performed to minimize fragments that are too small.

% Methylation	% Missing	Coverage Depth	Reads per Lane (M)
15	0	10	350

	Enzyme Pair	Fragments		Fragment loss			Reads per Sample	Samples per Lane	Reduction Rate
		Total	Selected	Methyl	Mut	Mis			
?	HincII - Sau3AI	1,010,764	495,970	74,396	65	0	4,215,090	83	14%
?	NsiI - BfaI	1,006,411	480,948	72,142	65	0	4,087,410	86	14%
?	HincII - BfaI	1,006,090	492,271	73,841	65	0	4,183,650	84	14%
?	HindIII - BfaI	976,772	448,559	67,284	65	0	3,812,100	92	13%
?	HindIII - Sau3AI	962,686	470,766	70,615	65	0	4,000,860	87	14%
?	NsiI - Sau3AI	954,889	498,360	74,754	65	0	4,235,410	83	15%
☑	HincII - MspI	636,098	243,241	36,486	65	0	2,066,900	169	8%
⊖	AccB1I - Sau3AI	604,746	289,477	43,422	65	0	2,459,900	142	8%
☑	HindIII - MspI	601,971	224,034	33,605	65	0	1,903,640	184	8%
⊖	AccB1I - BfaI	598,665	298,057	44,709	65	0	2,532,830	138	9%
?	NsiI - MspI	598,383	235,378	35,307	65	0	2,000,060	175	8%
⊖	AccB1I - MspI	474,622	155,393	23,309	65	0	1,320,190	265	5%
?	HincII - HinP1I	421,287	145,691	21,854	65	0	1,237,720	283	5%
?	NsiI - HinP1I	387,486	119,460	17,919	65	0	1,014,760	345	4%
?	HindIII - HinP1I	379,253	113,344	17,002	65	0	962,770	364	4%
⊖	AccB1I - HinP1I	352,448	87,965	13,195	65	0	747,050	469	3%
☑	PstI - Sau3AI	159,241	74,662	11,199	65	0	633,980	552	2%
?	PstI - BfaI	157,047	79,805	11,971	65	0	677,690	516	2%
?	PstI - MspI	138,967	43,220	6,483	65	0	366,720	954	1%
?	PstI - HinP1I	125,750	26,620	3,993	65	0	225,620	1551	1%

Table of Enzyme Pairs sorted by the number of selected fragments in the distribution plot.

Enzyme Pair
- ⊖ No Compatible buffer found
- ? No compatible buffer found; however, NEB is a supplier and is switching many of the enzymes to the CutSmart buffer
- ☑ Compatible buffers found

Fragments
- **Total** the total number of fragments between the enzyme pairs.
- **Selected** number of fragments between enzyme pairs and size selection.

Fragment loss
- **Methyl** Number of fragments expected to be lost due to methylation at a cutting site
- **Mut** Amount of fragment loss expected base on random mutation. https://doi.org/10.1016/j.tig.2010.05.003
- **Mis** Number of missing fragments allowed. (base on input above)

Reads per Sample
Number of reads needs to meet required coverage depth

Samples per Lane
Estimated number of samples that can fit on a lane given the expected reads per lane.

Reduction Rate
Percentage of the genome remaining after digestion and size selection

Fig. 2 A table showing the result of a typical analysis

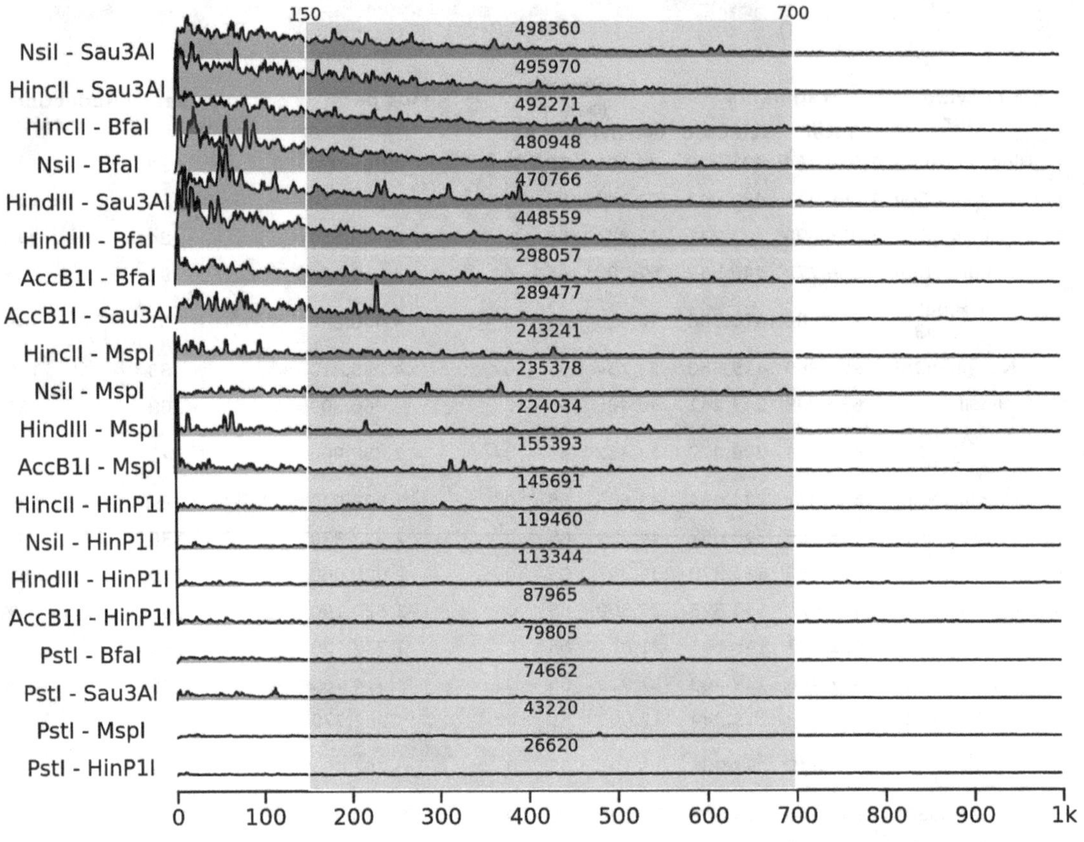

Fig. 3 Length distribution of fragments between the enzyme pairs. Size selection is in gray and sorted by the number of fragments in the selection area

11. Vortex the bead suspension every two samples while pipetting to maintain the beads in suspension so that the correct volume of beads is drawn each time.
12. When using magnetic beads for size selection, 4–6 samples may be processed each time while maintaining high yield with good purity.
13. If 96 samples are pooled, the resulting volume will be approximately 720 μL and 576 μL ($576 = 0.8 \times 720$) of the beads should be added. If less than 96 samples are pooled, the volume of bead suspension to be added should be equal to $0.8 \times$ (the total volume of the pooled samples).

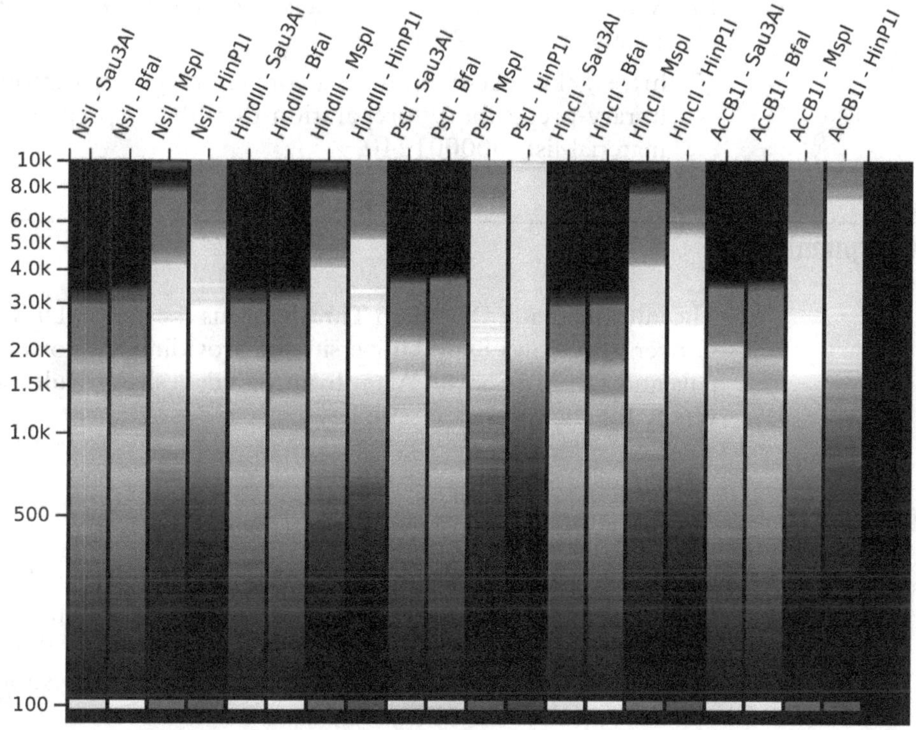

Fig. 4 Expected gel image of double-digestion products

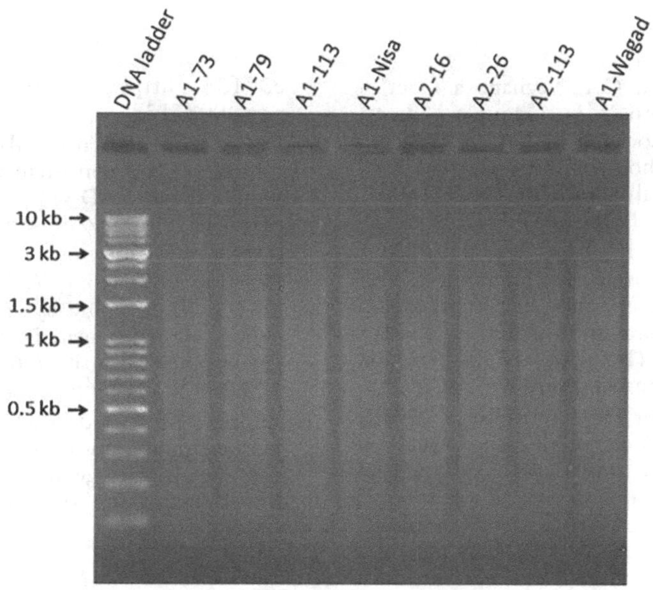

Fig. 5 Agarose gel image of fragment profiles after double digestion with *Hind*III (RCRE) and *Sau3*A1 (FCRE)

14. A total 2 mL assay working solution will be sufficient for 2 standards plus 7 samples.
15. https://knowledge.illumina.com/library-preparation/dna-library-prep/library-preparation-dna-library-prep-reference_material-list/000001240.

Acknowledgments

The authors thank Dr. Adam Thrash for his assistance, Dr. Corinne Grover from Iowa State University for providing the cotton materials and Dr. Md Samsul Alam from Bangladesh Agricultural University for the rohu carp samples.

Funding

This research was funded, in part, by USDA ARS awards [58-6066-0-066 and 58-6066-0064] to DGP. The USDA ARS was not involved in the design and conduct of the study nor in the analysis and interpretation of the data and in the decision to submit the results for publication.

Conflict of Interest The authors have no conflicts of interest to declare that are relevant to the content of this chapter.

References

1. Magbanua ZV, Hsu C-Y, Pechanova O et al (2023) Innovations in double digest restriction-site associated DNA sequencing (ddRAD-Seq) method for more efficient SNP identification. Anal Biochem 662:115001. https://doi.org/10.1016/j.ab.2022.115001
2. Arick MA, Grover CE, Hsu C-Y et al (2023) A high-quality chromosome-level genome assembly of rohu carp, Labeo rohita, and its utilization in SNP-based exploration of gene flow and sex determination. G3 (Bethesda) 13. https://doi.org/10.1093/g3journal/jkad009
3. Peterson BK, Weber JN, Kay EH et al (2012) Double digest RADseq: an inexpensive method for de novo SNP discovery and genotyping in model and non-model species. PLoS One 7: e37135. https://doi.org/10.1371/journal.pone.0037135
4. DaCosta JM, Sorenson MD (2014) Amplification biases and consistent recovery of loci in a double-digest RAD-seq protocol. PLoS One 9: e106713. https://doi.org/10.1371/journal.pone.0106713
5. Peterson GW, Dong Y, Horbach C, Fu Y-B (2014) Genotyping-by-sequencing for plant genetic diversity analysis: a lab guide for SNP genotyping. Diversity 6:665. https://doi.org/10.3390/d6040665
6. Andrews KR, Good JM, Miller MR et al (2016) Harnessing the power of RADseq for ecological and evolutionary genomics. Nat Rev Genet 17: 81–92. https://doi.org/10.1038/nrg.2015.28

Chapter 17

Fast and Accurate Enrichment of Canid Excrement for Species* and Other Analyses (FAECES*) Method for Noninvasive Genotyping of Field-Collected Canid Scat

Lillian D. Parker, Jesús E. Maldonado, and Michael G. Campana

Abstract

The FAECES protocol details a method for obtaining multi-locus SNP genotypes from noninvasively collected scat samples. It employs a capture enrichment technique that can be adapted for various vertebrate species. The workflow encompasses probe design using BaitsTools, DNA extraction using the Qiagen DNeasy PowerSoil Pro kit, genomic DNA library preparation via the Santa Cruz Reaction (SCR) protocol, in-solution hybridization capture with myBaits probes, and comprehensive bioinformatic analysis utilizing tools such as FastQC, TrimGalore, Nextflow, and SNPRelate. The protocol aims to differentiate between species, identify individuals, and determine sex from scat samples, emphasizing flexibility and adaptability for diverse research applications.

Key words SNP genotyping, Noninvasive sampling, Capture enrichment, Hybridization, Bioinformatics, Genotyping identification, Sex determination

1 Introduction

The objective of this protocol is to obtain multi-locus single nucleotide polymorphism (SNP) genotypes from noninvasively collected scat samples. The method was originally developed to differentiate between coyote and kit fox scats and identify individual coyotes and kit foxes and their sex in a single assay [1]. However, capture enrichment is highly flexible, and this protocol can be modified for any vertebrate species or loci of interest given sufficient reference data for probe design. The workflow described includes probe design, DNA extraction, genomic DNA library preparation, hybridization capture using RNA probes, and bioinformatic analysis (Fig. 1).

Anthony Torres (ed.), *Genotyping: Methods and Protocols*, Methods in Molecular Biology, vol. 2943,
https://doi.org/10.1007/978-1-0716-4642-7_17,

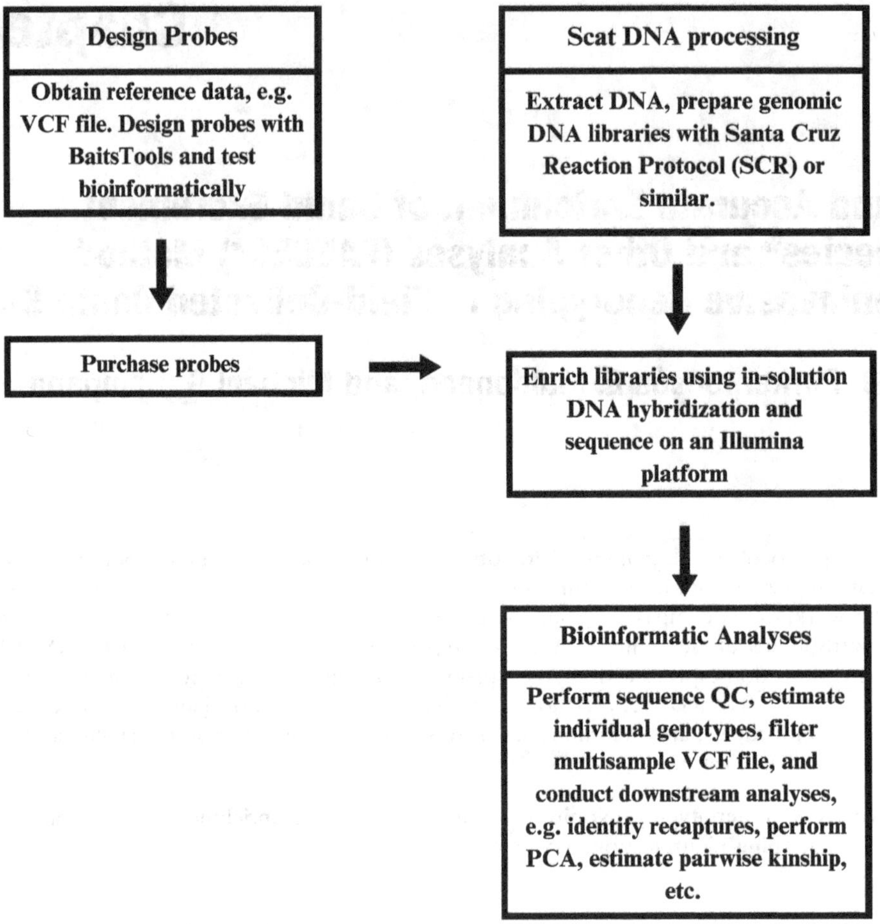

Fig. 1 Diagram of FAECES* workflow

2 Materials

2.1 Probe Design

1. **Input File**: Variant call format (VCF) file or other supported input file (FASTA alignment, PyRAD loci file, etc.).
2. **Software**: Installed BaitsTools [2] software: https://github.com/campanam/BaitsTools.

2.2 DNA Extraction

1. **Kit**: Qiagen DNeasy® PowerSoil® Pro kit.
2. **Equipment**:
   (a) Microcentrifuge.
   (b) Microcentrifuge tubes (2.0 and 1.5 mL).
   (c) Pipettor (1000 μL).
   (d) Vortex mixer.

2.3 Genomic DNA Library Preparation: The Santa Cruz Reaction (SCR) Protocol by Kapp et al. [3]

1. **Pipettors**: (10 and 100 μL).
2. **Qubit fluorometer**.
3. **DNA Shearing Method**: Bioruptor® Pico sonication device or other method.
4. **Thermocycler** (with heated lid).
5. **Magnetic particle collector** (for strip tubes).
6. **Microcentrifuge**.
7. **qPCR thermocycler**.
8. **Library visualization devices**: Agilent TapeStation or Bioanalyzer Systems.
9. **Cold rack** (for strip tubes).
10. **Oligonucleotides** [3]:
    - (a) SCR_P5_adapter: /5AmMC12/ ACACTCTTTCCCTACACGACGCTCTTCCGATCT.
    - (b) SCR_P7_adapter: /5Phos/ AGATCGGAAGAGCACACGTCTGAACTCCAGTCAC/3AmMO/.
    - (c) SCR_P5_splint: /5AmMC6/ NNNNNNNAGATCGGAAGAGCGTCGTGTAGGGAAAGAGTGT /3AmMO/.
    - (d) SCR_P7_splint: /5AmMC12/ GTGACTGGAGTTCAGACGTGTGCTCTTCCGATCTNNNNNNN /3AmMO/.
    - (e) qPCR primers IS7 and IS8 (as described in ref. [4]): IS7, 5′-ACACTCTTTCCCTACACGAC-3′; IS8, 5′-GTGACTGGAGTTCAGACGTGT-3′.
    - (f) Index primers (as described in ref. [5]).
    - (g) P7 and P5 Illumina primers (from myBaits® protocol, Arbor Biosciences): P5: 5′- AATGATACGGCGACCACCGA-3′, P7: 5′-CAAGCAGAAGACGGCATACGA-3′.
11. **Reagents**
    - (a) 1 M Tris–HCl (pH 8.0).
    - (b) 1 M $MgCl_2$ (pH 8.0).
    - (c) 10% Tween-20.
    - (d) 0.5 M EDTA (pH 8.0).
    - (e) 100% glycerol.
    - (f) 5 M NaCl.
    - (g) UltraPure water.
    - (h) T4 DNA ligase (2,000,000 U/mL, NEB M0202M).
    - (i) T4 PNK (10,000 U/mL).

(j) NEB (M0201L).
(k) ET SSB (500 ng/μL).
(l) NEB (M2401S).
(m) ATP (100 mM).
(n) Thermo Fisher (R0441).
(o) DTT (1 M, P2325) (*See* **Note 1**).
(p) T4 RNA ligase buffer (NEB, B0216L).
(q) KAPA SYBR® FAST qPCR Master Mix kit.
(r) BSA (NEB B9200).
(s) KAPA HiFi HotStart ReadyMix (or enzyme of choice for index PCR).
(t) SPRI beads (Thermo Fisher A58521) Recipe following [6].
(u) Ethanol (200 proof).
(v) Nuclease-free water (H_2O).
(w) Polyethylene glycol (PEG) 8000 50% w/v (*See* **Note 2**).

12. **Buffers**: Prepared following [3]
    (a) TE buffer: 10 mM Tris–HCl (pH 8.0), 1 mM EDTA (pH 8.0).
    (b) EBT buffer: 10 mM Tris–HCl (pH 8.0), 0.05% Tween-20.
    (c) Adapter dilution buffer: 1× T4 RNA Ligase Buffer, 0.05% Tween-20.
    (d) SSB dilution buffer: 20 mM Tris–HCl (pH 8.0), 20 mM NaCl, 0.5 mM DTT.
    (e) SCR buffer: 666 mM Tris–HCl, 132 mM $MgCl_2$.

2.4 In-Solution Hybridization Capture

1. **Kit**: myBaits® hybridization capture kit (Arbor Biosciences).
2. **Required Equipment** (not included in kit):
    (a) Plastics: 50 mL tubes, 1.5 mL microcentrifuge tubes, 0.2 mL PCR strips with attached lids.
    (b) Heat block (for microcentrifuge tubes).
    (c) Thermocycler (with heated lid).
    (d) Optional: Pippin Prep for size selection (Sage Science).

2.5 Bioinformatics Analysis

1. **Hardware**
    (a) A computer with a multicore processor with 64 GB of RAM and at least 1 TB of disk storage; individual needs will depend for the scale of the project.

2. **Programs and Tools**
   (a) Current versions of the following:
   (b) FastQC [7].
   (c) TrimGalore [8].
   (d) Nextflow [9].
   (e) Conda.
   (f) BWA-MEM [10].
   (g) SAMtools [11].
   (h) SNPRelate (R package) [12].
   (i) R environment (R Project for Statistical Computing, https://www.r-project.org/).
   (j) Elephant analysis Nextflow pipeline (https://github.com/campanam/Elephants) [13].
   (k) *RMP_calc.R* from mPCRselect (https://github.com/ellieearmstrong/mPCRselect) [14].
   (l) VCFtools [15].
   (m) MAFFT [16].
3. **Reference Genome**
   (a) A reference or pseudo-reference genome based on previous sequencing (e.g., RADseq) that has undergone repeat-masking.
4. **Preliminary Data**
   (a) Data used to design probes (e.g., FASTA sequence or alignment of sequences, VCF file, ipyrad loci file, etc.).

3 Methods

3.1 Probe Design Using BaitsTools Software (See Note 3)

1. **Install pipeline**: Follow installation instructions at https://github.com/campanam/BaitsTools.
2. **Mask low complexity and repetitive regions:**
   (a) Use software like **RepeatMasker** [17] or **GenMap** [18] to mask low complexity repetitive regions in your genome reference or pseudo-reference.
3. **Using VCF files for probe design:**
   (a) Use the subcommand vcf2baits if you are using a VCF file. List all subcommand-specific arguments by running "baitstools vcf2baits -h." Here we detail steps to design probes based on a VCF file including variants from two canid species aligned to the same reference genome (CanFam 3.1) [19].

4. **Recommendations for probe design targeting SNP loci for noninvasive applications**
   (a) **Probe length**: Use "-L 80" to specify 80 bp probes. This length provides sufficient specificity and accommodates hybridization with the shorter DNA molecules from noninvasive samples.
   (b) **Alternate alleles: Use** "-a" to generate probes with alternate alleles to minimize the potential for allelic dropout.
   (c) **Tiling**: Use "-O 20" to specify an offset of 20 bp. For 80 bp probes, this creates approximately 4× tiling of each variant site, increasing the probability of hybridization by increasing the number of probes covering each site.
   (d) **Masked bases**: Use "-K 25" to exclude baits with more than 25% masked bases, such as those in repetitive regions of the genome.
   (e) Options can be changed to suit each application.
5. **Balancing targeted variants across taxa:**
   (a) If targeting more than one taxon, we recommend balancing the number of targeted variants for each.
   (b) Define groups (populations or species) in a tab-separated value (TSV) file with sample name as it appears in the VCF file in the first column, and population assignment in the second.
   (c) Use "--taxacount" to specify the target number of variants that are variable across all taxa, homozygous within groups but variable across taxa, and variable within a subset of taxa.
   (d) Use "--popcategories" to specify a comma-separated list of the maximum number of population-specific variants in order of appearance in the taxa TSV file.
   (e) Example: In [1], 100 variants that were fixed in kit foxes and coyotes but differed between species were selected to differentiate between them. A maximum of 500 SNPs that were variable within each species was selected. Before setting this maximum, a greater number of SNPs were found for coyotes, and this option was necessary to balance the number of markers for each species.
6. **Filter VCF.**
   (a) Filter your VCF file to include only the selected variants using VCFtools [15], e.g., "vcftools --vcf <your.vcf> --bed <baits.bed> --out <filtered.vcf>."
7. **Downstream Analysis**: Perform downstream analysis to evaluate the performance of the selected loci for your future application.

   (a) PCA or pairwise kinship analysis with SNPRelate [12] or calculate the probability that two randomly selected individuals in a population will share a genotype (random match probability) using *RMP_calc.R* from mPCRselect [14].
   (b) *See* Subheading 3.5 Bioinformatic analysis for additional details.

8. **Sex Markers**
   (a) To determine sex, add additional probes targeting sex-specific genes if available, e.g., zinc finger (*ZFX/ZFY*) or amelogenin (*AMELX/AMELY*).
   (b) Use aln2baits or tilebaits in BaitsTools [2] with the same options as above (e.g., 80 bp probes with 20 bp offset and 4× tiling).
9. **Submit candidate probes for synthesis**
   (a) We have used myBaits® probes from Arbor Biosciences. Arbor Biosciences may suggest further filtration of selected loci.

3.2 DNA Extraction (Scat Samples)

1. **Sample Storage:**
   (a) Upon collection, store scat samples with silica desiccant in the field and freeze them until further processing to minimize DNA degradation.
2. **Subsampling Scat:**
   (a) Use sterile forceps, cleaning with 20% bleach and rinsing with water between samples.
   (b) For dehydrated samples, use 100 mg of material, for non-dehydrated sample use 200 mg.
   (c) Store subsamples in DNAse free, lo-bind 2.0 mL flip-top microcentrifuge tubes.
3. **DNA Extraction Protocol**
   (a) Follow the Qiagen DNeasy® PowerSoil® Pro protocol. If not targeting microbial DNA, omit bead beating. Instead, add 800 μL CD1 to the 2.0 mL tube with the subsample and vortex well.
4. **Incubation**
   (a) Incubate the subsamples overnight at 56 °C in a shaking incubator.
5. **Final Steps and Storage**
   (a) Continue with the protocol as written in the manual.
   (b) Elute in 75–100 μL of CD6 solution.
   (c) Store DNA at −20 °C until ready for library preparation.

3.3 Genomic DNA Library Preparation: The Santa Cruz Reaction (SCR) Protocol by Kapp et al. [3]

Here, we use the Santa Cruz Reaction (SCR) Protocol [3] for genomic DNA library preparation. This single-stranded DNA (ssDNA) method is optimized for ancient DNA applications and can accommodate a broad range of input DNA quantities from fewer than 3 ng (tested with as little as 0.93 ng) to more than 75 ng [3]. Scat-derived DNA samples vary widely in their quantity and quality, depending on time since defecation [20], the presence of moisture [21], temperature and UV exposure [22], diet [20, 23], etc. The flexibility of the SCR method accommodates this variability for efficient library preparation with minimal formation of adapter dimers. However, other commercial methods are available including ssDNA protocols (e.g., IDT xGen ssDNA and low-input DNA library preparation kit, and Claret Biosciences SRSLY kits) and dsDNA protocols (NEB Ultra II DNA Library Prep Kit). The SCR protocol is described in detail in the supplementary materials in [3]. Below, we briefly describe the steps involved in library preparation, highlighting those that differ from the protocol as originally written.

1. **Quantification**
   (a) Quantify extracted DNA using a Qubit fluorometer with a 1× dsDNA HS assay.
2. **Input DNA Based on Concentration Tiers**
   (a) Refer to Table S8 [3], reproduced below, to determine the appropriate input DNA tier. We have tested the SCR protocol with starting dsDNA up to 100 ng.

Tier	Picomoles ssDNA	dsDNA (ng)	P5 (μM)	P7 (μM)	SSB (ng/μL)
1	1.0–2.5	29–100	12	6	328
2	0.5–0.99	15–29	6	3	164
3	0.25–0.49	7–15	3	1.5	82
4	0.12–0.24	3–7	1.5	0.75	41
5	<0.12	<3	0.75	0.375	20.5

3. **DNA Shearing**
   (a) Shear DNA to an average fragment length of approximately 250 bp using your method of choice. Here, we use a Bioruptor® Pico sonication device.
   (b) If necessary, dilute DNA in water for a final volume of 50 μL with up to 200 ng of DNA. This will provide enough DNA for two library preparations. Remaining sheared DNA can be stored at −20 °C. Shearing is not part of the original protocol, which was developed for ancient DNA samples that are already sufficiently degraded.

4. **Buffer Preparation**
   (a) Prepare buffers, master mixes, adapter dilutions, SSB dilutions, and sheared DNA ahead of time to streamline the library preparation process. Prepare buffers including TE buffer, EBT buffer, adapter dilution buffer, SSB dilution buffer, and SCR buffer [3].
5. **Adapter Preparation**
   (a) Prepare SCR_P5 and SCR_P7 adapter mixes in 0.2 mL tubes. SCR_P5: 30.6 μL of water, 5 μL of 10× T4 RNA ligase buffer, 6 μL of 100 μM SCR_P5_adapter, 8.4 μL of 100 μM SCR_P5_splint. SCR_P7: 30.6 μL of water, 5 μL of 10× T4 RNA ligase buffer, 6 μL of 100 μM SCR_P7_adapter, 8.4 μL of 100 μM SCR_P7_splint.
   (b) Hybridize by first incubating the mixes at 95 °C for 1 min in a thermocycler with a heated lid (105 °C), then ramp down to 10 °C, decreasing 0.1 °C per second. Follow oligonucleotide quality control workflow (S10—S12 in the supplemental information) [3].
6. **Adapter and SSB Dilution**
   (a) Dilute adapters to the correct concentration depending on the tier of input DNA using Table S8 [3]. Use adapter dilution buffer to make dilutions. Dilutions should go through no more than five freeze/thaw cycles, so plan accordingly.
   (b) Dilute stock SSB using SSB dilution buffer to desired concentration for input DNA tier. Diluted SSB should be used within 1 month.
7. **Master Mix Preparation**
   (a) Prepare SCR master mix [3].

Reagent	Stock concentration	1× (μL)	Final concentration	20× (μL)
PEG 8000	50%	20	20%	400
SCR buffer	13.3×	3.75	1×	75
DTT	1 M	0.5	10 mM	10
ATP	100 mM	0.5	1 mM	10
T4 PNK	10,000 U/mL	0.625	0.125 U/mL	12.5
T4 DNA ligase	2,000,000 U/mL	0.625	25 U/mL	12.5

8. **Sample Preparation**
   (a) Prepare DNA samples in 0.2 mL strip tubes with attached caps. Add 20 μL of each sheared DNA sample. Prepare libraries in batches according to tier for ease of pipetting (e.g., prepare all tier 1 samples together, so that only tier 1 p5/p7 adapters and SSB are thawed).
9. **Library Preparation**
   (a) Follow the SCR protocol [3]. First, combine sheared DNA with SSB and heat to denature. Cool reactions, then combine with SCR master mix and adapters at the correct concentration for your input DNA, mix well, and incubate to ligate adapters.
10. **SPRI Bead Cleanup**
    (a) Proceed immediately to clean up with SPRI beads [3]. Because of the high concentration of PEG in the reaction, first add 55 μL EBT to dilute, then add 75 μL SPRI solution to each reaction and proceed with clean up [5]. Elute in 50 μL EBT.
11. **Library Storage**
    (a) The library can be stored at −20 °C or proceed directly to qPCR or index PCR.
12. **qPCR Optional**
    (a) Prior to indexing PCR, you may perform a qPCR to determine the number of amplification cycles for each sample. Here, we use Kapa SYBR®Fast Universal Master Mix. First, dilute libraries 1:100 with TET buffer. Assemble a reaction mix following the recipe below:

    1× reaction mix

    (a) Kapa SYBR®Master mix: 10 μL.
    (b) IS7 primer (10 μM): 0.04 μL.
    (c) IS8 primer (10 μM): 0.04 μL.
    (d) H_2O: 7.2 μL.
    (e) BSA (2.5 mg/μL): 1.0 μL.
    (f) Aliquot 19 μL of the reaction mix per sample to a qPCR plate. Add 1 μL of 1:100 diluted library and mix. Seal plate and cycle with the following conditions: 95 °C for 3 min, followed by 39 cycles of 95 °C for 3 s, 60 °C for 20 s, and 72 °C for 20 s, capturing fluorescence at the extension step.
13. **Index PCR Cycle Determination**
    (a) To determine the number of cycles of index PCR to perform, we recommend starting with *n* cycles = Cq −

5; i.e., subtracting 6 cycles from the Cq to account for the 1:100 library dilution and adding one to increase yield.

14. **Index PCR**

    Here, we use Kapa HiFi HotStart ReadyMix. Begin by making a reaction mix following the recipe below:

    1× reaction mix (multiply by the number of samples you are preparing):

    (a) Kapa HiFi HotStart ReadyMix: 12.5 μL.

    (b) BSA (2.5 mg/μL): 1.0 μL.

    (c) H_2O: 5.0 μL.

    (d) Aliquot 18.5 μL reaction mix per sample into strip tubes with attached caps.

    (e) Making sure to avoid cross contamination, carefully add 1.0 μL of a unique i7 primer (10 μM) and 1.0 μL of a unique i5 primer (10 μM) to each sample such that each sample has a unique combination.

    (f) Add 5.0 μL of library.

15. **Thermocycling Conditions**

    (a) Mix well and place on a thermocycler with a heated lid. Use the following cycling conditions and the number of cycles indicated by the qPCR (*n* cycles):

    (b) 98 °C for 45 s.

    (c) Followed by *n* cycles of 98 °C for 15 s, 60 °C for 30 s, and 72 °C for 1 min followed by a final extension of 72 °C for 5 min.

16. **SPRI Bead Cleanup and Library Quantification**

    (a) Cleanup reactions with SPRI beads with ca. 1.5× concentration of beads and elute in 30 μL nuclease-free water.

    (b) Quantify the indexed library using Qubit and visualize on a TapeStation, Bioanalyzer, or similar.

17. **Library Yield Adjustment**

    (a) Determine the total yield after index PCR and determine if the cycle number should be increased. For libraries with large proportions of exogenous (non-target) DNA such as those derived from scat, Arbor Biosciences recommends putting as much DNA as possible into capture, up to 12 μg. We suggest a minimum of 500 ng, aiming for 1–2 μg per sample. Repeat index PCR for a total of 3–6 PCR replicates depending on the yield from each reaction. Performing replicate PCRs increases the library complexity going into capture and helps to minimize sample failure and missing data downstream.

18. **Final Quantification**
    (a) Combine replicate index PCRs for each sample and clean the combined reactions with SPRI beads as written above. Elute in 25–30 μL nuclease-free water. Quantify using Qubit.

3.4 In-Solution Hybridization Capture and Sequencing

1. **Pool libraries or capture each separately**
   (a) Arbor Biosciences suggests pooling as few samples as possible. We have tested capture with one, two, and three samples in a single capture reaction. If combining multiple samples into a single reaction, perform an additional qPCR on the raw DNA extract to estimate the amount of endogenous DNA content and combine samples with similar quantities of endogenous DNA. If targeting nuclear DNA, choose a single copy nuclear gene (e.g., [24]).
2. **Choose a hybridization temperature**
   (a) Choose an appropriate temperature based on the expected divergence between the baits and target sequences (higher temperatures for increased stringency). For species-specific baits, we have used 62 °C (*See* **Note 4**). Program a thermocycler with a heated lid for the denaturing and hybridization process: 95 °C for 5 min, hybridization temperature for 5 min, followed by hybridization temperature forever. Lid temperature should be 5°–10° higher than each step temperature to minimize evaporation.
3. **Dry libraries**
   (a) Dry samples using a DNA vacuum centrifuge in 0.2 mL tubes with attached lids to prepare for capture. We recommend individual tubes rather than strip tubes. Libraries should be suspended in 7 μL water or nuclease-free buffer.
4. **Prepare the hybridization mix**
   (a) Follow the myBaits® manual specific to the baits used to prepare the hybridization master mix. Here, we describe methods based on v5.03. We follow the "High Sensitivity Protocol," but for our noninvasive samples, we have found efficient results with a single enrichment round by adding 5.5 μL baits per sample to the hybridization mix. Incubate the mix at 60 °C for 10 min, then at room temperature for 5 min. Aliquot 18.5 μL of reaction mix into strip tubes.
5. **Prepare the blockers mix**
   (a) Follow the myBaits® manual specific to your baits kit to prepare the blockers master mix. Add 5 μL of the mix directly to the libraries (at 7 μL) and mix by pipetting (total volume 12 μL).

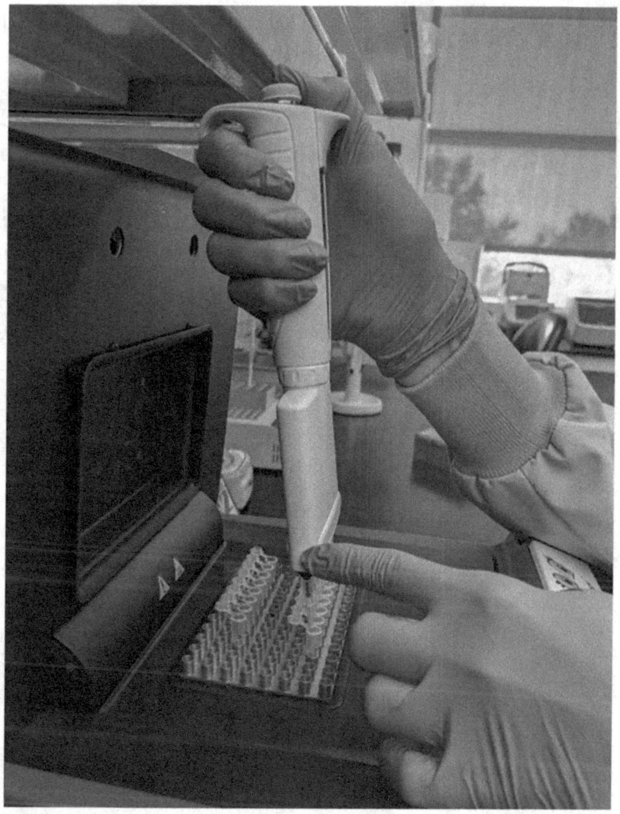

Fig. 2 Combining denatured libraries with hybridization mix on a thermocycler

6. **Hybridization set up**
   (a) Follow the myBaits® protocol for the hybridization set up procedure. Briefly, heat libraries + blocker mix to 95 °C for 5 min to denature, reduce to the hybridization temperature, and put baits mix on the thermocycler. Incubate for 5 min. Once the baits mix has reached the hybridization temperature, leaving the tubes on the thermocycler, pipette the baits mix to the libraries + blocker mix and gently mix by pipetting (Fig. 2). Briefly spin hybridization mix down and return to the thermocycler. Continue hybridization by incubating overnight, approximately 16–24 h.
7. **Recover captures**
   (a) Recover the captures by separating the bait-bound library from the other library molecules. Follow the myBaits® protocol specific to your kit. Briefly, captures are recovered by using streptavidin-coated magnetic beads to bind bait + target molecules. A magnetic particle separator is then used to separate the bead + bait + target complex from the rest of the library molecules. The target

molecules are washed with warm buffer, and most of the non-target DNA is washed away. At the bead washing step, the supernatant containing non-target DNA libraries can be saved and used for re-capture or capture with different baits. For example, you may wish to enrich first for nuclear targets as with the FAECES* probes, and then mitochondrial targets [25]. At the final wash, elute with 30 μL buffer included in myBaits® kit.

8. **Amplify captured libraries**
   (a) Amplify captured libraries with Illumina P5/P7 primers either while still bound to beads (if using Kapa HiFi HotStart or NEB Ultra II Q5 polymerases) or after first heating to separate beads and library molecules. Follow the instructions in the myBaits® protocol for cycling conditions for Kapa HiFi. qPCR can be performed to determine the cycle number. myBaits® recommends the fewest number of cycles possible to have sufficient yield for sequencing. Alternatively, you can amplify for 14 cycles and evaluate yield and perform a second PCR as needed. For noninvasive applications, 14–16 cycles are common.
9. **Quantify captured libraries**
   (a) After cleaning the PCR reaction with SPRI beads, quantify using Qubit and visualize with the TapeStation (or similar) to determine if adapter dimer is present and to record molarity.
10. **Pool samples for sequencing**
   (a) Combine samples equimolarly for sequencing on an Illumina platform. If dimer is present, estimate the concentration of library in the target sequencing range excluding dimer and use this value to make pooling calculations.
11. **Optional size selection**
   (a) Use a Pippin Prep (Sage Sciences, 1.5% or 2% agarose cassette) to size select libraries and remove adapter dimer after using the TapeStation images to determine the range to target (e.g., 180–500 bp). Visualize the final library pool using TapeStation and submit for sequencing. Some sequence cores offer Pippin prep size selection as a service if necessary.
12. **Estimate the number of reads needed**
   (a) If using FAECES* probes, we recommend estimating 4–6 million reads per sample. Although most SNPs are recoverable within the first two million mapped sequence reads, given the low cost of sequencing and the high variability in the percent of reads that map to the reference genome for scat samples [1], targeting more than two million reads is recommended.

3.5 Bioinformatic Analyses

1. **Sequence QC**
   (a) Upon receiving sequences, check their quality and quantity using FastQC [7].
2. **Trim Adapter Sequences**
   (a) Use TrimGalore [8] to remove adapter sequences. Use the "--illumina" flag to trim Illumina adapters and the "--paired" flag for read pair validation, i.e., if one of the reads in a pair is less than 20 bp after adapter removal, the pair is removed. Optionally, use "--fastqc" to run FastQC on trimmed output files.
3. **Run Elephants NextFlow Pipeline**
   (a) Create or edit a config file to prepare to run the Elephants Nextflow pipeline [13], which automates the processes of aligning sequences to a reference, processing sample alignments with SAMtools [11], and identifying variants using GATK [26]. Follow the instructions at https://github.com/campanam/Elephants.
   (b) Prepare a CSV file with sample information in the format "Sample,Library,Read1,Read2." This allows you to include multiple libraries from a single sample and combine into a single ".bam" alignment file.
   (c) Run the pipeline with the following command: nextflow run campanam/Elephants -r main –c <your.config.file.config>. Pipeline outputs .bam alignment files, individual variant files (gVCFs), and mapping statistics.
4. **Merge Individual Variant Files into a GVCF**
   (a) Merge the individual variant files into a single GVCF file using the "CombineGVCFs" tool from GATK [26]:

```
gatk CombineGVCFs \
-R ref.fasta \
--variant sample1.g.vcf.gz \
--variant sample2.g.vcf.gz \
-O population.g.vcf.gz
```

5. **Perform Joint Genotyping**
   (a) Perform joint genotyping on the output file from **step 4**, <population.g.vcf>, using the GATK tool "GenotypeGVCFs" [26]:

```
–java-options "Xmx4g" GenotypeGVCFs \
-R ref.fasta \
-V population.g.vcf.gz \
-O output.vcf.gz
```

6. **Filter the VCF File**
   (a) Use VCFtools [15] to filter the VCF from **step 5**. The filters that are used depend on the application. In Parker et al. (2022) [1], the following filters were used:
   - Indels were removed first ("--remove-indels").
   - Sites were then limited to those in the baited regions ("--bed").
   - Sites were limited to those with a minimum individual read depth of five ("--minDP 5").
   - All samples with >50% missing data were removed from the VCFfile (as determined by "--missing-indv").
   - Then, sites with >25% missing data ("--max-missing .75"), or with a minor allele count < 2 ("--mac 2") were removed.
   - Finally, one SNP per 500,000 bp was selected ("--thin 500,000").
7. **Conduct Downstream Analyses with Filtered VCF File**
   (a) For example, use SNPRelate [12] to perform principal components analysis (PCA). First, load the SNPRelate library and dependencies in your R environment.
   (b) Read the VCF file into R, e.g., vcf.fn <- ("in.vcf").
   (c) Convert to Genomic Data Structures (GDS) format with "snpgdsVCF2GDS."
   (d) Perform PCA with "snpgdsPCA." If following the FAECES* workflow, use PCA to differentiate between coyotes and kit foxes, separate the raw VCF file into two files by species, and refilter loci.
   (e) Using the respective single-species VCF files, you may also estimate pairwise kinship with maximum likelihood estimation using "snpgdsIBDMLE."
   (f) Identify recaptures of individuals of each species by calculating pairwise identity-by-state (IBS) with SNPRelate [12], function "snpgdsIBD."
8. **Estimate the Random Match Probability**
   (a) Calculate the random match probability (RMP), i.e., the probability that two randomly selected individuals will share a genotype using the *RMP_calc.R* script from the mPCRselect package [14].
   (b) The input for this script is an allele frequency file calculated with VCFtools ("vcftools --gzvcf $output.vcf.gz --freq --out out.frq").
   (c) The script is run by the following command: Rscript RMP_calc.R out.frq.

9. **Determine Sex**
   (a) Re-run the Elephants Nextflow pipeline [13] using the relevant female reference sequence (e.g., *ZFX* gene) and male reference sequence (e.g., *ZFY* gene) separately.
   (b) Identify sequence variants using SAMtools "mpileup" without the -C 50 option followed by bcftools call [1]. If using the FAECES* probe set, preliminary sex determinations can be assigned automatically from the *ZFX* alignment files using an ad-hoc script which assumes the Canfam3.1 reference [19] (available at: https://github.com/campanam/faeces).
   (c) Otherwise, make consensus sequences using "samtools consensus -o outfile -f fasta in.bam." Align *ZFX* and *ZFY* consensus sequences with reference sequences using MAFFT [16] and determine sex by eye.

4 Notes

1. DTT must be fresh. To avoid freeze store 1 M aliquots at −20 ° C. Unusually high Cq values may indicate DTT that is not fresh enough.
2. PEG should be stored at −20 °C and thawed in aliquots (e.g., 2 mL) to avoid excessive freeze/thaw cycles. If not stored properly, PEG may cause the reaction to fail.
3. Alternatively, mPCRselect (https://github.com/ellieearmstrong/mPCRselect) [14] may be used to select SNP loci and design probes for enrichment. This pipeline selects markers with the highest F_{st} between groups for population assignment and selects markers with the highest nucleotide diversity for individual identification.
4. Depending on the expected divergence between baits and target sequences, different hybridization temperatures can be used. Arbor Biosciences recommends 63 °C for bait-target divergence <15%, 60 °C for expected divergence 15–25%, and 55 °C for >25% divergence or DNA inserts mostly shorter than the probe length. Some experimentation may be necessary to optimize capture conditions, but in our experience, the reaction is robust to different conditions.

Conflict of Interest

The authors have no conflicts of interest to declare that are relevant to the content of this chapter.

References

1. Parker LD, Campana MG, Quinta JD et al (2022) An efficient method for simultaneous species, individual, and sex identification via in-solution single nucleotide polymorphism capture from low-quality scat samples. Mol Ecol Resour 22:1345–1361. https://doi.org/10.1111/1755-0998.13552
2. Campana MG (2018) BaitsTools: software for hybridization capture bait design. Mol Ecol Resour 18:356–361. https://doi.org/10.1111/1755-0998.12721
3. Kapp JD, Green RE, Shapiro B (2021) A fast and efficient single-stranded genomic library preparation method optimized for ancient DNA. J Hered 112:241–249. https://doi.org/10.1093/jhered/esab012
4. Gansauge M-T, Aximu-Petri A, Nagel S, Meyer M (2020) Manual and automated preparation of single-stranded DNA libraries for the sequencing of DNA from ancient biological remains and other sources of highly degraded DNA. Nat Protoc 15:2279–2300. https://doi.org/10.1038/s41596-020-0338-0
5. Meyer M, Kircher M (2010) Illumina sequencing library preparation for highly multiplexed target capture and sequencing. Cold Spring Harb Protoc 2010:pdb.prot5448. https://doi.org/10.1101/pdb.prot5448
6. Rohland N, Reich D (2012) Cost-effective, high-throughput DNA sequencing libraries for multiplexed target capture. Genome Res 22:939–946. https://doi.org/10.1101/gr.128124.111
7. Andrews S (2010) FastQC: a quality control tool for high throughput sequence data. http://www.bioinformatics.babraham.ac.uk/projects/fastqc
8. Krueger F (2019) Trim Galore. https://github.com/FelixKrueger/TrimGalore
9. Di Tommaso P, Chatzou M, Floden EW et al (2017) Nextflow enables reproducible computational workflows. Nat Biotechnol 35:316–319. https://doi.org/10.1038/nbt.3820
10. Li H (2013) Aligning sequence reads, clone sequences and assembly contigs with BWA-MEM. arXiv 1303.3997v1
11. Li H, Handsaker B, Wysoker A et al (2009) The sequence alignment/map format and SAMtools. Bioinformatics 25:2078–2079. https://doi.org/10.1093/bioinformatics/btp352
12. Zheng X, Levine D, Shen J et al (2012) A high-performance computing toolset for relatedness and principal component analysis of SNP data. Bioinformatics 28:3326–3328. https://doi.org/10.1093/bioinformatics/bts606
13. Prado NA, Armstrong EE, Brown JL et al (2023) Genomic resources for Asian (Elephas maximus) and African savannah elephant (Loxodonta africana) conservation and health research. J Hered 114:529–538. https://doi.org/10.1093/jhered/esad034
14. Armstrong EE, Li C, Solari KA, Mooney JA, Campana MG (2022–2024) mPCRselect. https://github.com/ellieearmstrong/mPCRselect
15. Danecek P, Auton A, Abecasis G et al (2011) The variant call format and VCFtools. Bioinformatics 27:2156–2158. https://doi.org/10.1093/bioinformatics/btr330
16. Katoh K (2005) MAFFT version 5: improvement in accuracy of multiple sequence alignment. Nucleic Acids Res 33(2):511. https://doi.org/10.1093/nar/gki198
17. Smit AFA, Hubley R, Green P. RepeatMasker, unpublished data. https://www.repeatmasker.org/
18. Pockrandt C, Alzamel M, Iliopoulos CS, Reinert K (2020) GenMap: ultra-fast computation of genome mappability. Bioinformatics 36:3687–3692. https://doi.org/10.1093/bioinformatics/btaa222
19. Hoeppner MP, Lundquist A, Pirun M et al (2014) An improved canine genome and a comprehensive catalogue of coding genes and non-coding transcripts. PLoS One 9:e91172. https://doi.org/10.1371/journal.pone.0091172
20. Panasci M, Ballard WB, Breck S et al (2011) Evaluation of fecal DNA preservation techniques and effects of sample age and diet on genotyping success. J Wildl Manag 75:1616–1624. https://doi.org/10.1002/jwmg.221
21. Regnaut S, Lucas FS, Fumagalli L (2006) DNA degradation in avian faecal samples and feasibility of non-invasive genetic studies of threatened capercaillie populations. Conserv Genet 7:449–453. https://doi.org/10.1007/s10592-005-9023-7
22. Vynne C, Baker MR, Breuer ZK, Wasser SK (2012) Factors influencing degradation of DNA and hormones in maned wolf scat. Anim Conserv 15:184–194. https://doi.org/10.1111/j.1469-1795.2011.00503.x
23. Murphy MA, Waits LP, Kendall KC (2003) The influence of diet on faecal DNA amplification and sex identification in brown bears (Ursus arctos). Mol Ecol 12:2261–2265.

https://doi.org/10.1046/j.1365-294X.2003.01863.x

24. McDonough MM, Parker LD, Rotzel McInerney N et al (2018) Performance of commonly requested destructive museum samples for mammalian genomic studies. J Mammal 99:789–802. https://doi.org/10.1093/jmammal/gyy080
25. Klunk J, Duggan AT, Redfern R et al (2019) Genetic resiliency and the Black Death: no apparent loss of mitogenomic diversity due to the Black Death in medieval London and Denmark. Am J Phys Anthropol 169:240–252. https://doi.org/10.1002/ajpa.23820
26. McKenna A, Hanna M, Banks E et al (2010) The Genome Analysis Toolkit: a MapReduce framework for analyzing next-generation DNA sequencing data. Genome Res 20:1297–1303. https://doi.org/10.1101/gr.107524.110

Index

Anthony Torres (ed.), *Genotyping: Methods and Protocols*, Methods in Molecular Biology, vol. 2943, https://doi.org/10.1007/978-1-0716-4642-7,